TIME

SPECIAL EDITION

The ExoMars module Schiaparelli separating and heading to Mars, expected to happen in October 2016

MISSION to MARS

Our Journey Continues

Contents

4   Introduction: The Race Is On

10   Buzz Aldrin: Over the Moon

THE JOURNEY

16   Live from Mars

24   A World Apart

26   A Brief History of Not Going

30   The Machines of Mars

36   Mission Possible

38   Neighbors in the Sky

THE PLAN

50   NASA's Next Steps

56   The White House Science Czar Gets Galactic

58   Mars on Earth

60   A Year in Space

66   How to Sneeze in Space

70   Rocket Business

THE ALLURE

80   Mars Art

86   The Red Muse

92   Will Man Outgrow the Earth?

94   Fact or Fiction: How Well Do You Know Mars?

An artist's concept of NASA's
planned Space Launch System

These Martian rock
formations suggest to
scientists that water once
flowed on the planet.

THE RACE IS ON
Mars has fascinated earthlings for thousands of years. Now we are closer than ever to visiting
By Jeffrey Kluger

I t was impossible to know if the beings on Mars were aware that we had found out they existed. There was no question at all that they were there, of course, and no question at all that they were highly intelligent. Had there ever been any debate, it was settled on the morning of Dec. 9, 1906.

It was a cold, snowy Sunday, at least in the East, and there was plenty of other news to occupy people's minds. There were whispers out of Europe, for example, that the celebrated Wright brothers were negotiating to license their lofty technology to aeronauts in Great Britain, accelerating the spread of flying machines to both sides of the globe. There was a kerfuffle in Washington, D.C., as Mark Twain, who was 71 and didn't care who he scandalized, came to town seeking access to the floor of the House of Representatives so he could lobby for a copyright bill. In New York, a former Utah senator was shot in his hotel room by his lover. According to that day's *New York Times*, a hotel maid discovered the woman standing over the wounded senator "attired in street costume, including her hat."

But it was the explosive, all-capital-letters headline elsewhere in the *Times*, running the full width of a page, that caused the real sensation. "THERE IS LIFE ON THE PLANET MARS," the headline announced. The news was simple, it was declarative, and the truth of what it asserted was impossible to deny.

No less an authority than Percival Lowell, the celebrated astronomer and the founder of the observatory in Flagstaff, Ariz., that bore his name, had made the discovery. His observations had revealed Mars to be crisscrossed with more than 350 canals, cut straight as

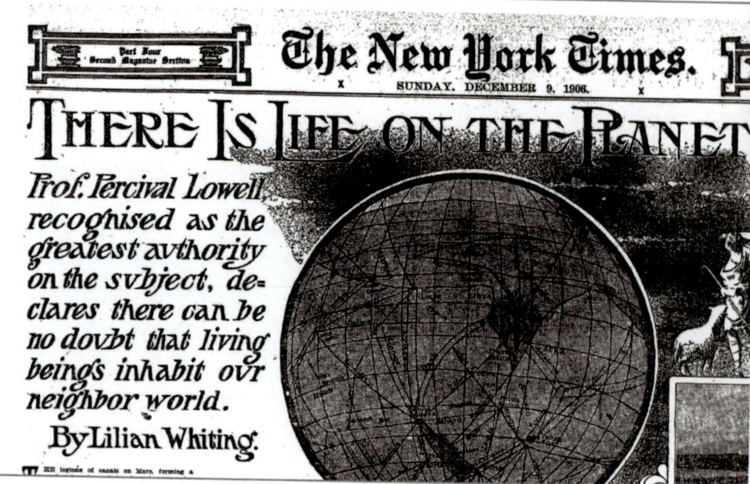

On Dec. 9, 1906, the *New York Times* announced the breaking news about Mars in a banner headline.

a ruler in sometimes-parallel lines—"like the twin rails of a railway," Lowell described them—and sometimes intersecting lines, all of them ultimately running to the snowcaps in the north and south poles. As the snows melted in the Martian spring and summer, water flowed all over the planet, hydrating the red desert and leading to a "vegetable quickening" that bloomed green-blue in the warm months and turned brown in the cold months, just as plants do on Earth.

Such a cunning network of clean, geometric viaducts could not be a natural phenomenon. "A thing made predicates a maker," as the *Times* put it. The canals were "an unanswerable argument for the existence of conscious, intelligent life."

History would eventually note that the ca-

nals were nothing of the kind. Far from proving anything, they were little more than the product of the tired eyes and wishful fantasies of a celebrated astronomer who, his genius aside, was still a human being, and human beings often see the things they dearly want to see.

History would go on to note that 59 years later, in the summer of 1965, that same *New York Times* would report that NASA's Mariner 4 spacecraft had swung by Mars, returning the first picture of the planet taken from so close, and discovered it to be a dry, broken, pitted world, little better than Earth's ruined moon. "THE DEAD PLANET," the now-wiser newspaper headlined its sorrowful editorial reporting the news. "The red planet is not only a planet without life now, but probably always has been." And so it seemed—not that that mattered to our imaginations and fancies.

THE FACT IS, HUMAN BEINGS HAD TUMBLED for Mars thousands of years before, and nothing was going to change that. The hold the planet had on us was deeper and more primal than anything a cold bit of contemporary science could shake loose.

Our fascination was partly fearful. Mars was the place that the monsters came from; the aliens in H.G. Wells's captivating 1890s novel *War of the Worlds* were the product not of the cool, blue clouds of Neptune or the great, lovely bulk of Jupiter but of the blood-red, up-close world next door. Venus was named after the goddess of love and Neptune after the god of the sea. Pluto might have been named for the god of the underworld, but that was more because of the deep blackness of the deep region in space it inhabits,

not because of any innate sinister traits.

But Mars was the Fire Star to the ancient Chinese and the Red One to the ancient Greeks, both of which sound faintly ominous. It was the Romans and their god of war who gave the planet the name that stuck. As for its two moons? They would be named Phobos and Deimos—for fear and panic. Nobody was meant to feel good about a planet like that.

Yet we did. Mars flirts with us, swinging wide of our planet at some points in its orbit as it passes around the opposite side of the sun, a quarter-billion miles away. And then it swings close, brushing by us once every two years at a distance of as little as 34 million miles—barely a long-distance call by cosmic standards. During those close approaches, it is the biggest and brightest object in the sky, save for the sun and the moon. It's at those moments that Mars has shown us its true nature, especially as telescopes have grown better and their resolution sharper.

Lowell might have been wrong about the water on Mars and wrong about the plants on Mars, and he was certainly wrong about the minds on Mars. But he wasn't wrong about the ice caps, which do grow and shrink and do contain water. He wasn't wrong about the seasonal color changes over the rest of the planet, though they're due to dust storms, not crops. Dust, though, implies atmosphere, and atmosphere implies . . . well, who knows what, but combined with even a little water, it must mean something.

There was something else that gave Mars promise. Jupiter, Saturn, Uranus and Neptune are the bruisers of the solar system, more than 10 times the diameter of Earth in Jupiter's case. But they're worlds without substance—or at least worlds without a surface—giant gas balls that you couldn't stand on so much as simply fall into. That's not the kind of place that seems likely to support life, even if it could emerge there. Mercury, Venus, Earth and Mars, on the other hand, are rocky worlds—solid worlds, worlds with a mantle and a crust that afford an organism a little purchase.

In the case of Earth and Mars, there's another thing too: both of them orbit the sun in what is known as the habitable zone or, more colloquially, the Goldilocks zone—the not-too-hot, not-too-cold region where liquid water can exist and organic molecules can do the combining and recombining they need to do to get biology started.

Against all that, scientific skepticism over Lowell's fever-dream canals or a single bad picture from Mariner 4 were nothing, and no sooner did humanity develop the ability to build big rockets than we began using them to throw spaceships at Mars. Only two robot probes have ever visited Mercury. Just one has gone to Uranus and Neptune; just one has gone to Pluto—and we didn't get out there until the summer of 2015.

Mars has been an entirely different matter. Since 1960, five space agencies—American, Russian, European, Japanese and Indian—have fired 43 different probes the Red Planet's way. Some never even made it to Earth orbit. Others failed en route to Mars or crashed into its surface. Plenty of others, however, have gotten there.

Now seven active probes are either orbiting Mars or crawling across its surface—a network of earthly machines that count as nothing short of interplanetary infrastructure.

When a spacecraft on the surface needs to communicate with Earth, it can toss its signal up to one passing overhead, which then flips the transmission home. When a comet speeding toward the sun passes Mars, controllers on Earth can instruct the orbiting probes to point their cameras in the right direction and report what they see. Our robotic human proxies aren't just studying Mars and working on Mars; they are, in a very real sense, inhabiting Mars.

MARS, IN TURN, IS REPAYING OUR ATTENtion. The planet may be every bit the cold, dry desert that the first bleak Mariner image suggested it was. But it may not have always been that way. Images returned by orbiters have shown a relief map of a world that's awash in water—except without the water. There are sinewy riverbeds, sprawling deltas, delicate alluvial fans, vast seabeds and ocean basins. Water—a great, great deal of water—surely once filled those basins and cut those riverbeds, even if there's little trace of it today.

NASA rovers that have scuffed in the soil and sampled its chemistry have confirmed that idea. The Martian surface is crusted with salts and other minerals that on Earth form only—or at least mostly—in the presence of water.

What's more, the planet's very wet era lasted a long time. For a billion or so of Mars's 4.5 billion years, scientists now believe, it had every bit the garden-world potential of Earth, with a thick atmosphere protecting a warm, hydrated surface, rich in organic chemicals. The planet's lack of a magnetic field, however, allowed solar wind to claw away the atmosphere from the top down. The planet's low

gravity—less than 40% that of Earth—made things worse, as every meteor hit that Mars sustained blasted more and more of its air into space. The result is an atmosphere that is today just 1% as dense as it once was. As the air bled away, the water evaporated and followed.

A billion watery years, though, is a long time—longer than it took the first primitive organisms to emerge on Earth. If life happened here, it might seem, it could possibly have happened there, and if it did happen there, that theoretical Martian life—resourceful and adaptable as all life is—may have retreated underground when the planet dried out, hunkering down in lingering reservoirs, where it may survive and thrive today. In 2015, NASA's Mars Reconnaissance Orbiter added weight to this idea when it discovered seasonal streaks of water flowing down Martian slopes, clearly the signs of those reservoirs warming and rising during the planet's spring and summer and retracting and retreating in fall and winter.

Mars, in our current thinking, is a heartbreak world, a planet that came very close to greatness and then caught all the wrong breaks at all the wrong moments. For the human species, though, that's enough. The existence of a neighboring planet with even a flickering chance of harboring life is irresistible in a cosmos that, until we know better, is insensible and dead except for the single biological beachhead that is Earth.

More than ever, then, the push for Mars is on. The same doggedness, the same cosmic monomania that once animated our drive for the moon seems to have found a new, more ambitious target. Robot probes continue to line up on the launchpads and sail off for Mars every two years, whenever the planets come into the conjunction that makes the trip the briefest and easiest. Where machines go, humans are increasingly trying to follow.

NASA has firmly fixed Mars as its next major goal for astronauts, with a new booster resembling the old Saturn V moon rocket and a new crew vehicle resembling the old Apollo spacecraft being developed and assembled. Money—increasingly, if slowly—is flowing from Washington to bankroll the effort. Private companies, such as California-based SpaceX—without the bottomless pockets of government, yes, but also without the politics that can make each round of funding such a fight—are working to jump the queue, building their own spacecraft, training their own crews and hoping to light out for Mars before the official space agencies can get out of their hangars.

It may all be folly. Mars is a monstrously hard place to reach and would be a far harder place to live. There's a reason that, if living things were there at all, they largely or entirely quit the place. Yet the improbability of any journey has rarely been an obstacle to humans trying to embark on it, especially when the destination is so intoxicating. Mars has been calling us since the days it was the Fire Star, the Red One—and it is calling us still. Now, at last, we may have the ability to answer.

> **WATER—A GREAT DEAL OF WATER—SURELY ONCE FILLED THE BASINS AND CUT THE RIVERBEDS.**

OVER THE MOON

Take it from a lunar pioneer: a manned mission to Mars will need some ambitious planning

By Buzz Aldrin

We have a rendezvous with destiny, a bold, audacious mission before us that will require a courage and resolve similar in kind to that which led to the first boot prints being planted on the moon 47 years ago. We were lifted by innovation and teamwork en route to that transformative, history-making lunar event that I was privileged to be part of, and those are the values we will need as we reach for the Red Planet—Mars.

Dispatching humans to walk upon Mars's dusty, eons-old landscape will be momentous and perilous, and building an ongoing presence there would naturally be an even more daunting prospect. Yet that should be our ultimate aim. To me, humans can undertake no greater endeavor than to create a permanent presence on another planet. I object to making a humans-to-Mars program a look-alike to our Apollo moon project: put people on the surface of the planet, proclaim success, have the crew set up experiments, plant a flag and then bring the explorers back to Earth. We can do more than that.

It has become clear to me that there's a universal hunger for a strong, vibrant and cutting-edge space agenda for the 21st century. And a drive to occupy Mars is a force that can unite the world's great nations in a purely peace-

Apollo 11 lunar module pilot Buzz Aldrin was the second man on the moon.

NASA
ALDRIN
UNITED STATES

From left: Aldrin walking on the moon in 1969; visiting with students at Aldrin Elementary School in Reston, Va., in 2015.

ful way. Several countries have long since launched robotic spacecraft that have orbited, landed on and even wheeled across Mars. With every new image relayed to Earth—specifically, from NASA's Opportunity and Curiosity rovers—there is, across our globe, a growing comfort level and even familiarity with a planet that is 45 million miles away. Martian landscapes strike a deep chord, perhaps because they have their doubles here on Earth.

I've devoted the past 30 years to studying Mars and to helping blueprint what is needed to establish and sustain a permanent human presence there. The newly formalized Aldrin Space Institute at the Florida Institute of Technology is working on shaping our strategies toward that end.

My plan to get humans to Mars employs a concept I call "Cycling Pathways for Exploration." The essential elements of the idea involve a progression of facilities as we move through areas of space. Think of them as a succession of service stations along a galactic highway, from low-Earth orbit (LEO) to cislunar space (between the Earth and moon) and then outward toward the moon, asteroids, Venus and Phobos—the inner moon of Mars. From there we would be in position to initiate a growing permanence of humans on the surface of the Red Planet.

IN KICK-STARTING OUR RENEWED SPACE voyages, there are a few things we need to do, a lot of which involve my old stomping grounds, the moon. We must organize a global LEO-lunar coalition to help us best implement and utilize the important features of what is known as "lunar reusable rendezvous." This involves parking habitat modules near the moon so that crews en route to Mars can pick up those modules and use them as extra living space for the long journey ahead. This builds upon my early work on rendezvous techniques—that is, the practice of bringing two or more ships together in space—that bolstered the step-by-step successes in America's Gemini and Apollo pro-

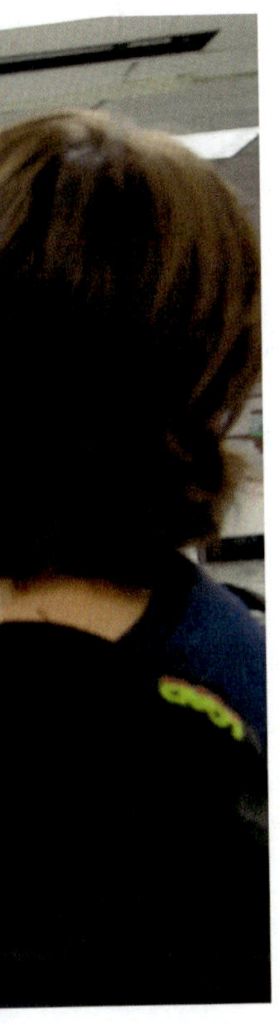

grams. I'm proud that my doctoral work at MIT helped refine techniques for NASA and led to John Hobolt's epic strategy of a lunar orbit rendezvous, which was so crucial in enabling us to get to the moon.

At the Aldrin Space Institute, work is ongoing to detail moon-support activities. We are designing equipment and methods with an eye toward the telerobotic assembly of all the facilities of the lunar base complex, including the crucial surface refueling depot that will convert −482°F ice to water, to hydrogen and oxygen, to liquid rocket fuel.

Another important lunar activity can prepare us with confidence for missions to the Red Planet. This one, though, does not involve *our* moon, but rather one of Mars's, Phobos. I dub the idea "pre-Phobos polar lunar orbit." Phobos is an easy landing site. It's small and has only light gravity. Getting to Mars from Phobos, which is orbiting the planet, will take practice, however. We should rehearse it first with a spacecraft orbiting Phobos that would send smaller modules to Phobos's surface before sending modules to Mars.

Collectively—through an international collaboration of technology—we can have a clear evolutionary plan for humans to permanently inhabit Mars. If we start now, I believe our outward reach can lead to Red Planet residence by 2040. That would be seven decades after the Apollo 11 moon landing, which occurred six and a half decades after the Wright brothers took flight at Kitty Hawk.

THROUGHOUT THE YEARS, I HAVE STUDIED cycling orbits between Earth and Mars. Cycling spaceships can shuttle back and forth between the orbits of the two planets, needing only minor adjustments in their trajectories on each cycle. This concept makes regular treks feasible and can, I believe, secure Mars as a second home for humanity.

On Mars, there's room to grow well beyond an initial outpost for expeditionary crews. There can be an expansion of standardized and interconnected modules to accommodate more people and more countries. Mars is a plentiful planet that has exploitable resources—most important, water—to sustain arriving explorers. Water on Mars is transformative. Along with other uses, we can harvest and process water for breathable oxygen, use it to maintain crops and use it as the basis of fuel for Mars space and land vehicles.

Nearly 50 locations on Mars have been advanced by scientists as potential landing spots for humans. Selecting a site depends on how safe the landing would be, how easy it would be to carry out operations from the site, and the ability of crews at the site to perform science experiments and duties while allowing for astronaut outreach to local resources.

As Yogi Berra—a presence in my hometown of Montclair, N.J.—once said, "It's tough to make predictions, especially about the future." There's ultimately no telling what lies ahead for Mars exploration, especially because settling there is not just a technical challenge but a managerial and political challenge as well. Ideas such as those I have advanced depend greatly on affordability. Moreover, I believe there's a need to reappraise the role of NASA: its goals and duties and the functions of its various field centers. Transitioning NASA from an operational agency to one that is largely advisory to other agencies may help ensure that it is noncompetitive with the zeal and spark of private industry and the blossoming of international rocket-launch capacity.

Now is the time to venture outward again, this time much farther into space. Since taking my first steps on the "magnificent desolation" of the moon those many years ago, I have worked to guarantee that those steps were not in vain and to advocate for strong American leadership in human spaceflight.

It is now that we must prove ourselves equal to the mission ahead. We shouldn't compete with other nations but rather should assist them to occupy the moon and to move farther on to the surface of Mars, continuing our legacy of exploration in peace for all mankind.

Apollo 11 moon walker Buzz Aldrin is an international advocate of space science and planetary exploration. Aldrin co-authored, with Leonard David, Mission to Mars: My Vision for Space Exploration *(2013) and, with Marianne Dyson, a children's book,* Welcome to Mars: Making a Home on the Red Planet *(2014). Aldrin and Ken Abraham's book* No Dream Is Too High: Life Lessons from a Man Who Walked on the Moon *was published in April 2016.*

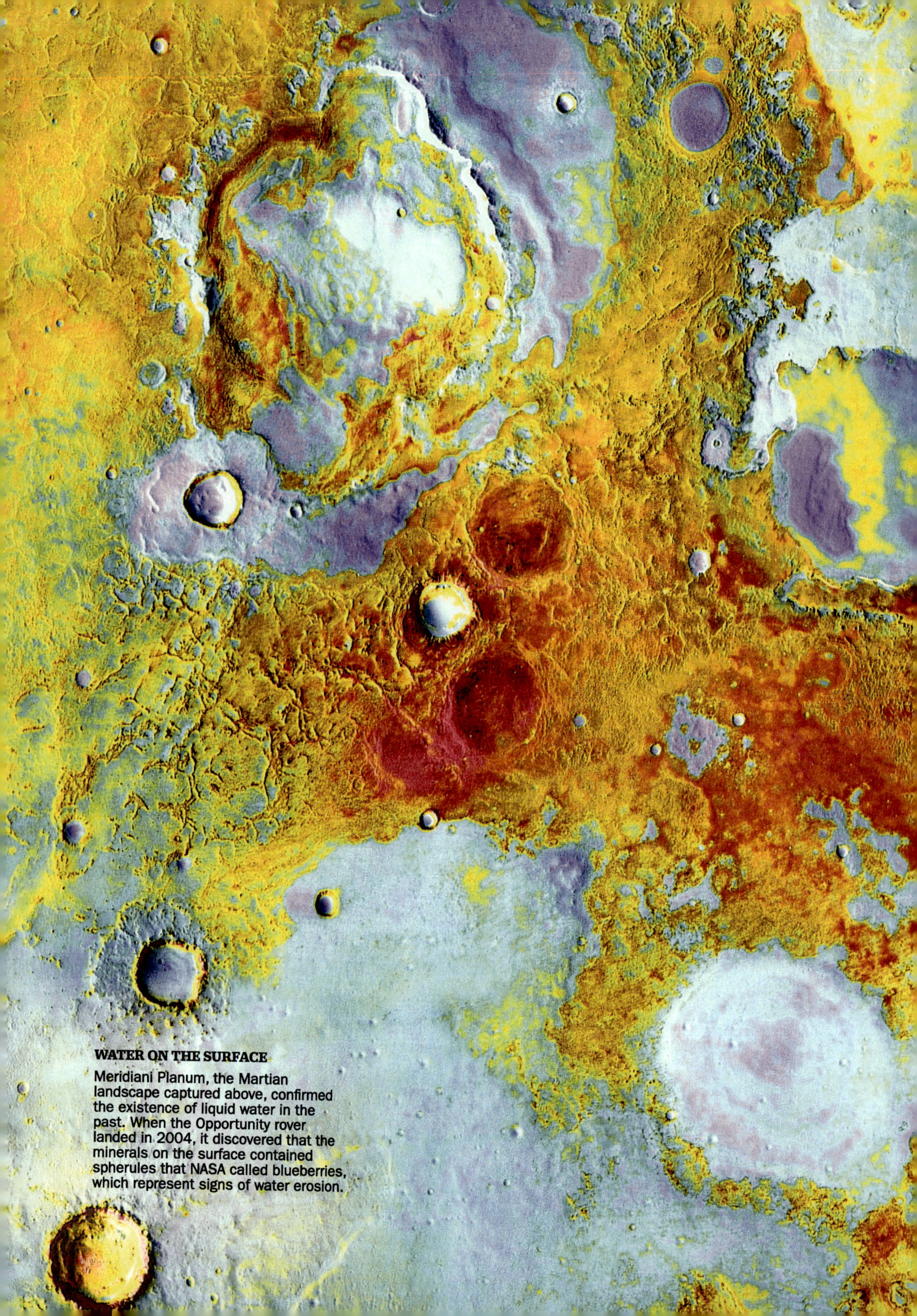

WATER ON THE SURFACE

Meridiani Planum, the Martian landscape captured above, confirmed the existence of liquid water in the past. When the Opportunity rover landed in 2004, it discovered that the minerals on the surface contained spherules that NASA called blueberries, which represent signs of water erosion.

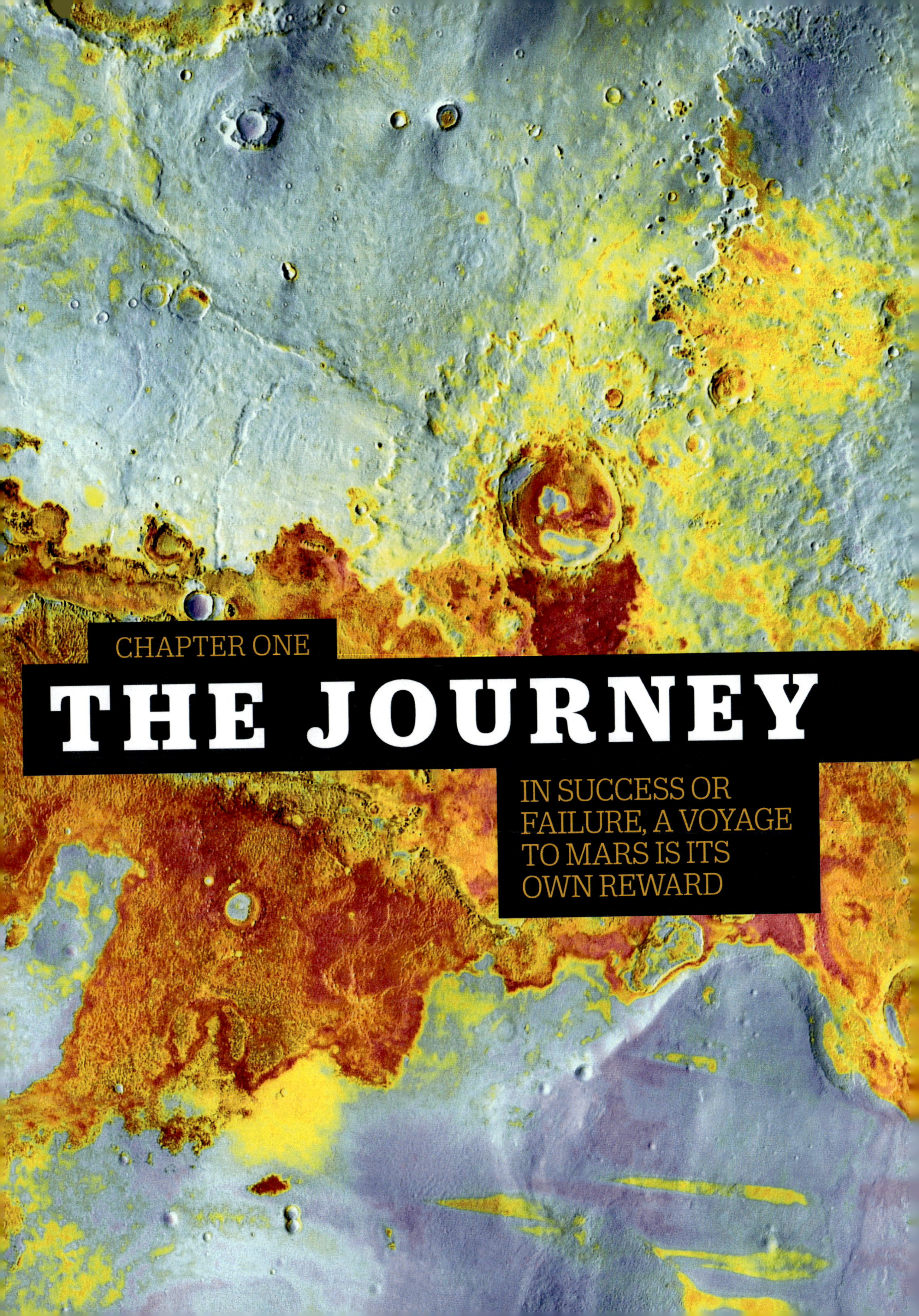
CHAPTER ONE
THE JOURNEY
IN SUCCESS OR FAILURE, A VOYAGE TO MARS IS ITS OWN REWARD

LIVE FROM MARS

A curious one-ton rover can teach us a great deal about the surface of the Red Planet

By Jeffrey Kluger

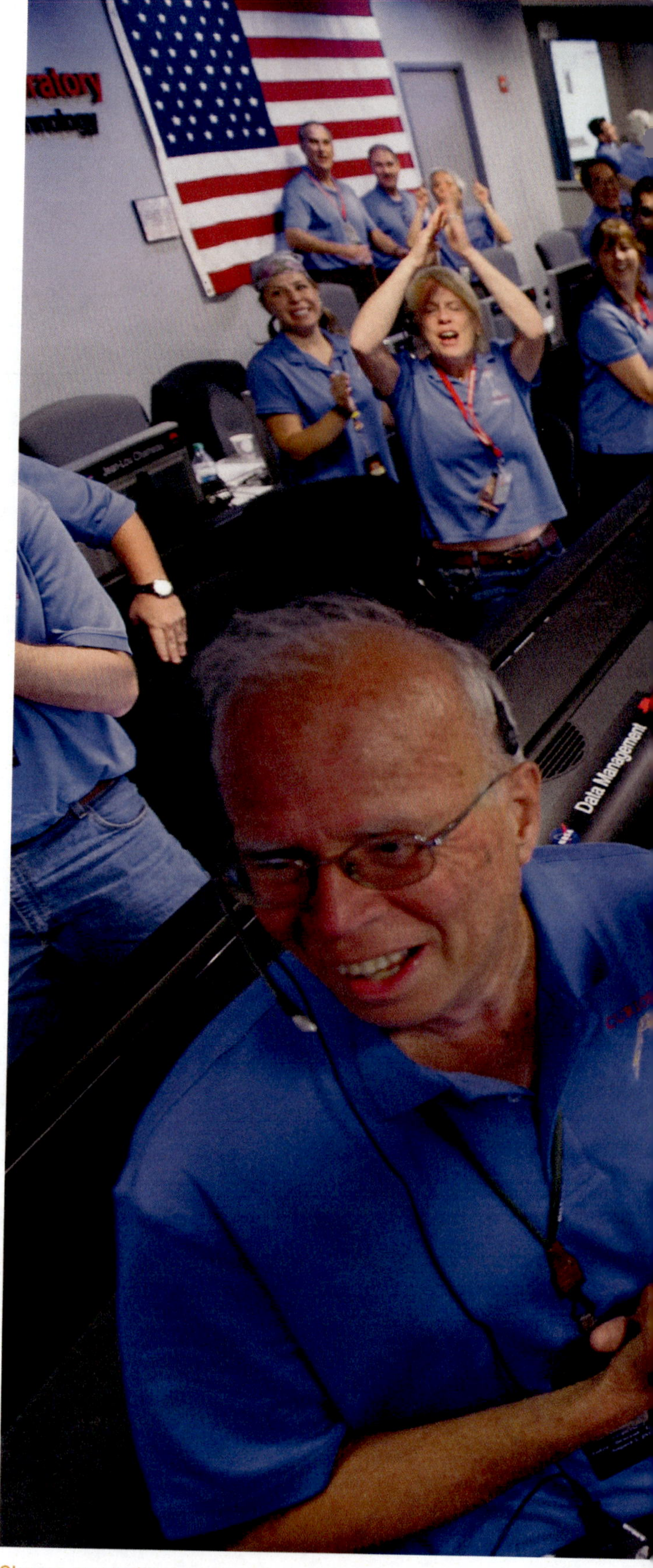

Cheers erupted among the team at NASA's Mars Science Laboratory on Aug. 5, 2012, when the Curiosity rover landed on Mars.

T he folks in mission control at NASA's Jet Propulsion Laboratory ate a lot of peanuts in the minutes leading up to the landing of the Curiosity rover on Mars in the summer of 2012. Peanuts have been the order of the day at JPL when a spacecraft is preparing to land ever since July 31, 1964, when the Ranger 7 probe was making its final approach to the moon. The Ranger's job was simple: to crash-land on the lunar surface, on the way down snapping a few thousand pictures to beam back home. Still, six previous Rangers had failed, and the JPL engineers knew they were about out of chances. Ranger 7 at last broke that losing streak, and as it happened, someone was nibbling peanuts during the landing. That, the missile men of JPL figured, was a powerful good-luck symbol—and no one has dared defy the tradition since.

Yet it would take more than luck and peanuts to get the SUV-size Curiosity rover safely to the surface of Mars. Sealed inside a blunt-bottomed capsule, the six-wheeled, $2.5 billion vehicle would slam into the Martian atmosphere at a blazing 13,000 mph. When it

was seven miles above the surface and the thin air had slowed the ship to 900 mph, its heat shield would pop away and it would deploy a billowing parachute. Its retrorockets would then bring the rover and its housing to a near hover just two stories above the surface, where it would be lowered to the ground by wire cables—an improbable extraterrestrial marionette, settling its wheels gently into the red soil. In Chicago, the Adler Planetarium held a late-night pajama party so families could follow the landing live. In New York, crowds gathered in Times Square to watch on a giant screen that usually shows only ads. NASA live-streamed the event, and the online traffic was so great—with up to 23 million people watching in the four hours immediately surrounding the landing—that its servers crashed.

What the people watching the live feed saw was not a spacecraft approaching Mars but a roomful of controllers in matching blue shirts, muttering about data acquisition and imager activation and drogue deployment and more. While much of that was incomprehensible, it was clear that something good was building. And then flight-dynamics engineer Allen Chen called, "Stand by for sky crane," and the room fell silent. Less than a minute later, he announced, "Touchdown confirmed! We're safe on Mars!" And with that, the silence was broken—explosively. "That rocked!" exclaimed then–deputy project manager Richard Cook as he took the stage at the celebratory press conference that followed. "Seriously, was that cool or what?"

It was cool indeed, but it was much more too. In an era in which the grind and gridlock of Washington have made citizens wary of anything the government touches, this was a reminder of what the country can do. The scene in mission control was what smart looks like. It was what vision looks like. Retrorockets could have eased Curiosity straight down to the surface, but that would have stirred up too much dust, perhaps fouling its works before it even got started. So the engineers chose the hard and creative and dangerous solution for the simple reason that it was also the best one.

A country that can't always get its roads

and bridges fixed at home has nothing short of a network of research stations on Mars. Two NASA orbiters—Mars Global Surveyor and Mars Odyssey—helped relay Curiosity's transmissions to Earth and wave in their newcoming sister for her landing. And even as Curiosity settled down to work, no fewer than eight other NASA probes were ranging through the solar system, exploring—or on their way to explore—the moon, Mercury, Jupiter, Saturn, Pluto, the asteroid Ceres and the interstellar void beyond the planets.

It's Curiosity, however, that may be the state of the exploratory art. With 10 instruments weighing a collective 15 times as much

In some ways, they've already done that, by framing an unavoidable question: If we can do this exceedingly hard thing so well, why do we make such a hash of the challenges at home, the inventing and investing that 21st-century progress demands? Help answer that one, and Curiosity could achieve great things on two worlds at once.

MARS MAY BE A METEOR-BLASTED DESERT today, but it was once a very different place. Its surface is marked with dry riverbeds, empty sea basins and even dusty oceans. Strip away 99% of Earth's atmosphere and boil off all its water, and it would look a lot like its desiccated cousin. Curiosity has found salts and other minerals on the surface that form principally in the presence of water, further confirming the planet's aquatic past. Planetary scientists now believe that Mars was wet for a billion of its 4.5 billion years, and as the early Earth proved, that is enough time to cook up life.

Curiosity's landing site is a formation known as Gale Crater, 96 miles wide. Located in the southern Martian hemisphere, it is thought to be up to 3.8 billion years old—well within Mars's likely wet period and thus once a large lake. A three-mile-high peak known as Mount Sharp rises in its center, with exposed strata layer-caked down its sides. Channels that appear to have been carved by water run down both the crater walls and the mountain base, and an alluvial fan—the radiating channels that define Earthly deltas—is stamped into the soil near the prime landing site. All of this is irresistible to geologists searching for the basic conditions for life.

Curiosity has conducted its work in numerous ways. The rover's arm scoops samples of soil and delivers them to an onboard analysis chamber, where they are studied by a gas chromatograph, a mass spectrometer and a laser spectrometer, looking for telltale isotopes, gases and elements. Chemical sniffers sample the Martian air for carbon compounds—especially methane—which are the building

as those aboard the earlier golf-cart-size rovers Spirit and Opportunity, Curiosity has spent the past four years, of what could be a decade or more of work, studying the geology, chemistry and possible biology of Mars, looking for signs of carbon, methane and other organic fingerprints on a world that a few billion years ago was warm and fairly sloshing with water. The previous rovers and landers had strongly made the case that Martian life—either extant or ancient—is possible, teeing Curiosity up to seal the deal. "We all feel a sense of pressure to do something profound," says geologist and project scientist John Grotzinger.

Curiosity uses its
long arm (not shown)
to take a selfie on
Oct. 31, 2012.

blocks and by-products of life. Martian geology is studied with a long-distance laser that can blast a million-watt beam at rocks up to 23 feet away, vaporizing them and allowing a spectrometer to analyze the chemistry of the residue. An onboard x-ray spectrometer does similar work on rocks that are closer to the rover. "With x-ray diffraction, we can really nail down what kind of mineral is there and how those rocks have formed," says deputy project scientist Joy Crisp.

Most appealing for the folks back home are the 17 cameras arrayed around Curiosity. They have the visual acuity to resolve an object the size of a golf ball 27 yards away and the resolution to capture one-megapixel color images from multiple perspectives. The sharpest of these imagers is mounted atop the rover's vertical mast, which, extended, rises seven feet aboveground. "You could not look this thing in the eye unless you were an NBA player," says mission systems manager Mike Watkins.

That camera, along with the other 16, has returned albums full of images that are in many ways even more important than the scientific findings the spacecraft sends home. Yes, the pictures are often an important part of those findings, allowing geologists on Earth to scrutinize the Martian surface as if they were crouching down on the dry, red soil themselves. But they serve another function too. Each sweeping portrait of an alien world that so incongruously resembles the United States' desert Southwest, and each image that Curiosity takes of itself—a terrestrial machine in a very non-terrestrial place—reminds us of the ambition and creativity that made a mission like this possible.

THE IMPULSE TO GET LOST IN THAT dreamy feeling and even to sentimentalize Curiosity—to treat it almost like a human astronaut—is hard to resist. "The rover is getting ready to wake up for its first day in a new place," said mission manager Jennifer Trosper at an early post-landing news conference. Describing what the science team's work schedule would be like, Watkins said, "The rover's day ends on Mars around 3 or 4 p.m. The rover tells us what she did today, and that . . . lets us plan her day tomorrow."

Such anthropomorphizing has always been present in space exploration in a way it isn't with other scientific endeavors. The confirmation of the elementary particle the Higgs boson earlier in the same summer that Curiosity landed was in some ways a much bigger development. But few people—outside the physics community, at least—sentimentalized it. Nobody calls a particle "she."

President Obama—like every president from Kennedy through the second Bush—was quick to make hay of good news from space. The mission, he said in a post-landing statement, "proves that even the longest odds are no match for [America's] unique blend of ingenuity and determination." And NASA administrator Charles Bolden—like every NASA chief who preceded him—was quick to give props to the president who appointed him. "President Obama has laid out a bold vision for sending humans to Mars in the mid-2030s," he said, "and today's landing marks a significant step in achieving this goal."

That kind of inspirational talk has continued to flow from both the administrator and the administration, and with the space-industry state of Florida very much in play in most elections, it's likely to continue from politicians well into the future. Obama's record on space has been mixed. The idea of privatizing the business of getting cargo and astronauts to low-Earth orbit raised eyebrows at first, but the move has been looking a lot smarter since Elon Musk's SpaceX Corp. and the Virginia-based Orbital Sciences began tag-teaming the job of sending unmanned resupply missions to the International Space Station. SpaceX and the Boeing Corp. are expected to begin shuttling astronauts up and down as early as 2017. Still, it's not easy to say exactly *how* private that private effort has been. NASA shared some of the R&D costs with the participating companies and signed lucrative contracts with them before they even proved they were up to the job—to the tune of more than $4 billion covered by taxpayers.

The president's plan got less clear and less credible when it came to human travel to deep space. NASA is developing a crew vehicle called Orion—essentially a souped-up Apollo spacecraft—and a heavy-lift booster dubbed the Space Launch System (SLS), similar to the venerable Saturn V. Returning to the old model of the expendable booster with the crew vehicle perched on top is a safe and smart decision after the disasters of the shuttle era, but that old model was well-funded. The first Saturn V was launched in 1967, the 13th and last in 1973, and nine of those rockets took people to the moon.

The SLS, which in one form or another has been in the planning stage since 2004, is not scheduled for its first crewed flight until 2023. After that, it would fly every other year—at best. It's not clear what its destination would be either.

The administration also spent years pushing a so-called asteroid redirect mission, which involved an unmanned craft finding a small asteroid and towing it to the vicinity of the moon, and astronauts then landing on it. The idea found few believers, and most analysts expect it to die quietly in the next president's administration. That leaves Mars, albeit on a vague timeline. Musk has boasted that he will have an unmanned spacecraft on Mars and will beat NASA's goal of landing astronauts there by a decade.

"This is a pace that doesn't make any sense," says John Logsdon, a professor emeritus at George Washington University's Space Policy Institute, of the government's schedules. "When Kennedy said he'd get to the moon by the end of the decade, he actually meant 1967, and he thought he'd still be president."

Kennedy, of course, wasn't hamstrung by the same kind of budget issues. "I would be thrilled if we could land on Mars in the 2030s, and I truly believe that is within the capability of this country," says John Grunsfeld, the former head of NASA's science mission directorate and a five-time shuttle astronaut. "I don't believe it is necessarily within the capability of this country with a flat budget."

What any nation can achieve and afford is, of course, at least partly a function of what it chooses to achieve and afford, even in straitened circumstances. The genius of Kennedy's

commitment to a lunar landing before 1970 was its simplicity: a single goal and a deadline. The current plan, with ever-changing destinations and dates, has none of that New Frontier clarity.

Kennedy's agenda was also advanced by other presidents and a cooperative Congress. The space push spanned four administrations—counting that of Eisenhower, who created NASA—and six Congresses. And while they often scrapped over the budget, they agreed on the goal. A legislature that can barely keep the Federal Aviation Administration funded is not an easy partner for any White House with grand ambitions. Space isn't free, but with NASA's budget hovering in the vicinity of just $18 billion per year, or 0.51% of the total federal budget, it's hardly a bankbreaker either. The Department of Defense, in contrast, gets $598 billion, or 16.9%. What's more, as with Defense, NASA research pays dividends. The Curiosity program has employed 7,000 high-tech workers across most of the 50 states. And as Grunsfeld has pointed out, the rover's chemical sniffers—sensitive to individual organic molecules—could have national-security applications at ports and airports.

But the extraordinary success of the Mars Curiosity rover masks a far greater truth about space exploration: it requires monomaniacal commitment and an exceedingly high tolerance for failure. In their own way, the JPL peanuts are a reminder of that fact. It's unimaginable in today's attention-deficit political climate that there would ever have been a Ranger 7 after the repeated failures of Rangers 1 through 6. But it took mastering unmanned crash landings before we could master unmanned soft landings. And it took mastering unmanned soft landings before Neil Armstrong—five years almost to the day after Ranger 7 made its suicide plunge into the moon's Sea of Clouds—could set his boot onto the Sea of Tranquility. That's the way science progresses: incrementally, patiently and ultimately spectacularly. Some of America's grandest moments have come when we've trusted that fact.

SCIENCE PROGRESSES INCREMENTALLY, PATIENTLY AND ULTIMATELY SPECTACULARLY.

Celebration ensues at the Spacecraft Control Center in Bangalore as India's orbiter enters Mars's orbit.

INDIA TAKES FLIGHT

New to the Mars scene, the nation got it right on its maiden voyage

G etting to Mars is hard, unless of course you're India, in which case it's apparently very easy. The U.S. failed the first time it attempted a robotic Mars mission, back in 1964. The old Soviet Union failed the first 10 times it tried. Japan has tried once and failed once, in 1998; the European Space Agency half failed—getting a spacecraft in orbit in 2003 but losing the intended lander.

But India? In its first start out of the gate, the country got things right, following a perfect launch on Nov. 5, 2013, with a perfect arrival in orbit around the Red Planet 10 months later, on Sept. 24, 2014.

The spacecraft, delightfully known as MOM—for Mars Orbiter Mission—owes its successful journey to a number of things. First was some deft flying. MOM got to Earth's orbit just minutes after it was launched but waited more than three weeks before it finally peeled out for Mars, as engineers at the Indian Space Research Organization (ISRO) burned its engine a total of five times to raise its altitude gradually. That conserved fuel and made the final kick out of orbit and toward Mars easier and more efficient.

The ISRO also made it a point to keep things simple. No one pretends that MOM is the most capable spacecraft ever sent to Mars. It has a camera and four other instruments for studying the atmosphere and the surface, and that's about it.

Simple, however, means fewer things to break down—and it also means cheap. The spacecraft cost the equivalent of $74 million. The movie *Gravity*, released the month before MOM launched, cost $100 million. The U.S., with five active spacecraft currently orbiting Mars or on its surface, remains the leader in Red Planet research. But India—the rookie—is now firmly in the game.

A WORLD APART

Mars is more similar to Earth than to anything else in the solar system, but it is a far less hospitable planet for human life

SIZE Earth has 3.5 times as much surface as Mars, but the planets have the same amount of dry land.

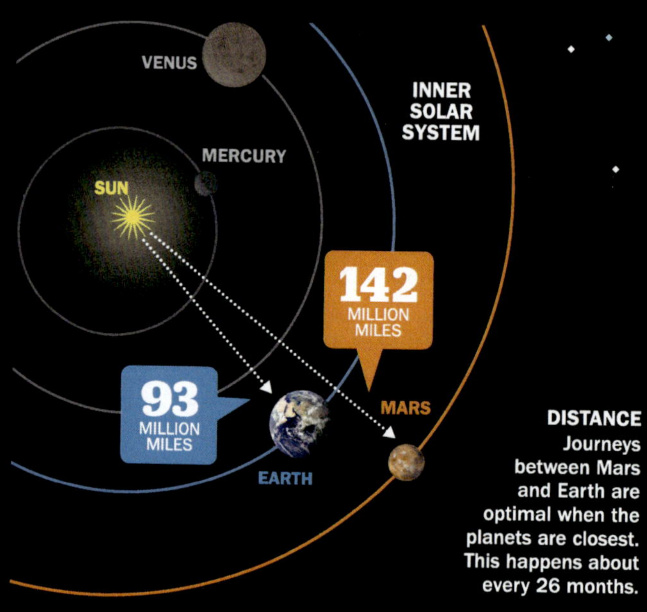

DISTANCE Journeys between Mars and Earth are optimal when the planets are closest. This happens about every 26 months.

A LONG JOURNEY

If Earth were the size of a quarter at the goal line of a football field, the moon would be about the size of a pencil eraser inside the one-yard line…

A MARTIAN YEAR

Mars completes a revolution around the sun in 687 days. A **30-year-old** on Earth would have had...

MARS'S MOONS

Mars has two tiny moons. Phobos orbits Mars in just seven hours—much faster than Earth's moon, which takes 27 days to orbit our globe.

TEMPERATURE RANGES

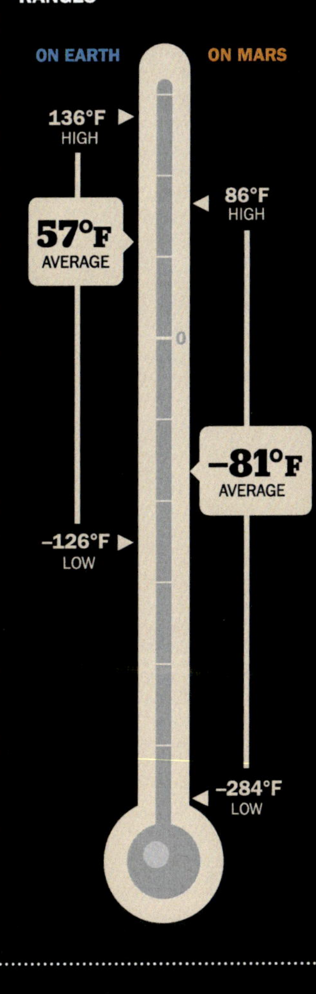

THE RED PLANET

Mars's red color comes from rusted iron in its soil. Even the sky has a red tint because dust in the atmosphere absorbs blue light and scatters red light.

WHERE'S THE WATER?

Along with water ice at the northern and southern polar caps, there is also evidence that briny water containing salts and other minerals flows near Mars's surface.

ATMOSPHERE

Mars's atmosphere is so thin—1% the density of Earth's—that hurricane wind speeds would feel like a light breeze.

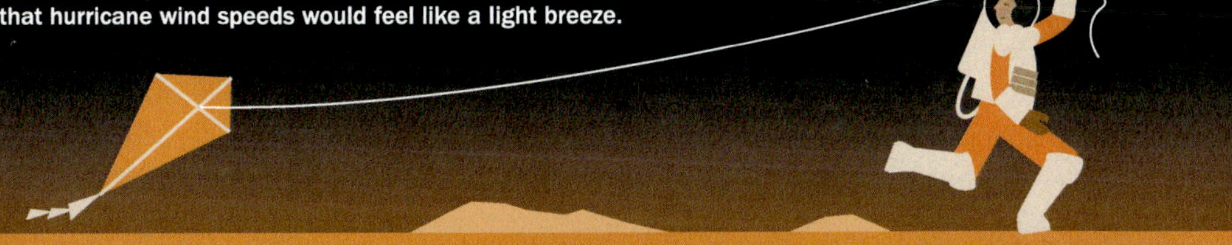

... Mars would be about the size of a dime in the opposite end zone—one reason travel between the real planets takes about eight months.

A BRIEF HISTORY OF NOT GOING

There have been many times we thought we would get to Mars in the near future. One of these days, we'll be right

By Lily Rothman

Whether imagined as a private-spaceflight plan or outlined as a NASA venture or suggested in a movie such as 2015's *The Martian*, the idea of a mission to Mars still seems, even in our technologically advanced world, like a thing of the distant future. Yes, such a mission would necessarily be futuristic—after all, it hasn't happened yet. But the concept is anything but. It's been decades since we earthlings first heard that reaching Mars was imminent.

Humanity had been interacting with Mars in literature since the dawn of modern science fiction (H.G. Wells's *The War of the Worlds* brought Martians to us in 1898), but it was in the mid–20th century that the idea of humankind going to Mars was first significantly discussed by scientists. Perhaps not coincidentally, that period also saw a wave of science fiction (stories like Robert Heinlein's *Red Planet* and Isaac Asimov's *The Martian Way*) that tended to take us there, rather than bringing them here.

Credit for this shift often goes to Wernher von Braun, a German-born rocket engineer who had worked with the Nazi regime before being brought to the U.S. near the end of World War II. In the early 1950s, by which point he was working with the U.S. Army on ballistic technology, von Braun published a book known in English as *The Mars Project*. Though he originally intended the book to be an explanatory appendix to a novel, its science and math were plausible enough to jump-start serious consideration of the idea.

In further publications over the next few years, von Braun expanded on his ideas. One component of his reasoning may seem quaint today: with war a thing of the past and all mankind living together in peace, he reasoned, we would be able to spare the jet fuel for such a trip. By the time von Braun wrote on the topic for a 1965 issue of *Popular Science*, the unmanned probes of the Mariner project had brought Mars further within reach, and he was able to confidently declare that "1986 wouldn't be a bad year" for such a mission to launch. (The year 1986 was a popular one for such projections, as the alignment of Earth and Mars would be more favorable at that point than at other close approaches.)

German-born American rocket scientist Wernher von Braun, left, and President John F. Kennedy tour Cape Canaveral on Nov. 16, 1963.

S-IV
UNITED STATES

Even as NASA geared up for a manned mission to the moon, its scientists were working on applying the same ideas to a manned mission to Mars. By 1966, TIME was reporting that possible jobs for the newest class of NASA astronauts would include just such a trip. And when the moon landing was a success on July 20, 1969, the following week's magazine contained an article with this tantalizing headline: NEXT, MARS AND BEYOND.

In that story, Vice President Spiro Agnew tempered expectations: though von Braun was still looking at the 1980s, Agnew thought the year 2000 was realistic. That September, a task force led by Agnew offered yet another timeline, suggesting that if NASA didn't get funding, "the Mars landing will not take place until after 1990." NASA, meanwhile, began to make plans. By 1972—listen to this, Hollywood—the agency even began wrestling with the question of what to do about astronauts' inevitable romantic entanglements during a 590-day mission.

Come 1985, by which time it was long clear that mankind would not reach Mars by 1986 or any year close to that, space experts across the globe floated the idea of a joint U.S.-Soviet mission. It was to be a Cold War–defying act of scientific greatness, made possible by splitting the bill. The imagined mission date: 2010. The notion was temporarily dashed after the *Challenger* space shuttle disaster in 1985, but a few years after that, dreams were revived. A July 18, 1988, TIME cover story about the progress the U.S. and the U.S.S.R. were making noted, "A manned trip to Mars, long the stuff of science fiction, now appears to be just a matter of time. The mystic planet, glowing red and ever brighter in the night skies, is heading toward its closest approach to the earth in 17 years this September, tantalizingly near and beckoning."

By the start of the 1990s, President George H.W. Bush had adjusted the target date with his Space Exploration Initiative, opening the door to aim for "astronauts on the red sands of Mars by 2019." Many experts agreed that sometime

around 2020 was realistic—but as the next decade progressed and other priorities intervened, the consistent news of probes and rovers going to Mars never led to headlines about concrete plans for a human being to follow.

At the dawn of the 21st century, George W. Bush announced that Mars was once again a top concern for NASA, promising that the inevitable discoveries along the way meant that "our efforts will be repaid many times over." The summer before the 2004 presidential election, he laid out a plan for returning to the moon around 2015, the date projected as the earliest possible timing, and getting to Mars not long after. The year 2030 frequently came up in reporting about Bush's Martian hopes, but the president—perhaps having learned from the decades of unfulfilled aspirations—never declared a firm deadline. Nonetheless, some space travel experts at the time saw 2030 as unnecessarily far away; with real dedication to the mission and a better program of funding, they estimated, something around 2014 could still be possible.

DESPITE ALL THE MISSED DEADLINES AND SCIENTIFIC DISAPPOINTMENT, HOPE REMAINS.

Now 2014 has come and gone, and 2030, in NASA time anyway, is practically around the corner.

But we haven't given up. Today, private spaceflight companies and national space agencies alike once again have Mars in their sights. The 2010 U.S. National Space Policy has a goal of getting humans to Mars orbit by the 2030s, and companies like SpaceX are developing new technology to help get them there. Experts know more than ever about what such a journey would entail—for example, the year astronaut Scott Kelly spent aboard the International Space Station represents the closest anyone has yet come to putting in the space hours that would be needed—and have once again dedicated themselves to the mission. Our collective record of staying closer to home might make some pessimistic about humanity's chances of reaching the Red Planet soon. Yet it's clear that, despite all the missed deadlines and scientific disappointment, hope remains. We're human, from the planet Earth, so we just can't help it.

SIX DECADES OF TRYING

A timeline of
the big idea

1952 WERNHER VON BRAUN'S BOOK *DAS MARSPROJEKT* DEBUTS IN WEST GERMANY.

1956 WITH WRITER WILLY LEY (RIGHT), VON BRAUN RELEASES *THE EXPLORATION OF MARS*, ARGUING THAT EXISTING TECHNOLOGY IS ENOUGH TO GET TO MARS.

1960 THE FIRST SOVIET ATTEMPT TO LAUNCH AN UNMANNED MISSION TO MARS BREAKS APART BEFORE MAKING IT OUT OF ORBIT.

1962 ERNST STUHLINGER OF NASA'S MARSHALL SPACE FLIGHT CENTER DESIGNS A PLAN FOR A MANNED MISSION IN THE EARLY 1980S.

1969 JEROME WIESNER, PRESIDENT KENNEDY'S SCIENCE ADVISER, SAYS, "IT WOULD BE A MISTAKE TO COMMIT $100 BILLION TO A MANNED MARS LANDING WHEN WE HAVE PROBLEMS GETTING FROM BOSTON TO NEW YORK CITY."

1965 IN JULY, NASA'S MARINER 4 ACHIEVES THE FIRST SUCCESSFUL FLYBY OF MARS.

1969 AFTER THE MOON LANDING ON JULY 20, SCIENTISTS AND POLITICIANS LOOK TO MARS AS THE NEXT BIG GOAL FOR THE U.S. SPACE PROGRAM.

1976 NASA'S VIKING 1 AND VIKING 2 LANDERS MAKE IT TO MARS AND SEND BACK HIGH-RESOLUTION IMAGES OF THE PLANET'S SURFACE.

1985 ASTRONOMER CARL SAGAN, ASTRONAUT SALLY RIDE AND SENATOR SPARK MATSUNAGA PROPOSE THAT THE U.S. AND THE U.S.S.R. WORK TOGETHER ON A PLAN FOR MARS.

1983 ACCORDING TO A PRESIDENTIAL TASK FORCE IN THE 1960S, WE COULD HAVE BEEN ON MARS BY THIS DATE IF CONGRESS HAD GIVEN NASA AT LEAST $8 BILLION A YEAR SINCE THE MOON LANDING.

1988 SOVIET OFFICIALS SAY THEY HOPE THE JOINT U.S.-U.S.S.R. MANNED MISSION WILL BE LAUNCHED BY 2010 AT THE LATEST.

1989 PRESIDENT GEORGE H.W. BUSH CALLS FOR A MANNED MISSION TO THE RED PLANET.

2004 AFTER A DECADE OF JUST PUTTING TECH ON MARS, PRESIDENT GEORGE W. BUSH MAKES IT A PRIORITY TO GET ASTRONAUTS THERE.

2015 *THE MARTIAN*, STARRING MATT DAMON, IS RELEASED AND RECEIVES SEVEN OSCAR NOMINATIONS.

2030s THE DECADE IN WHICH NASA BELIEVES IT WILL SEND A HUMAN TO MARS.

"

THE MACHINES OF MARS

If not for the rovers, landers and orbiters that have journeyed into space over the past several decades, we would know much less about our red neighbor. Here's a look at some of the machines, working or not, that played a part in the most important Mars missions

By Lisa Eadicicco

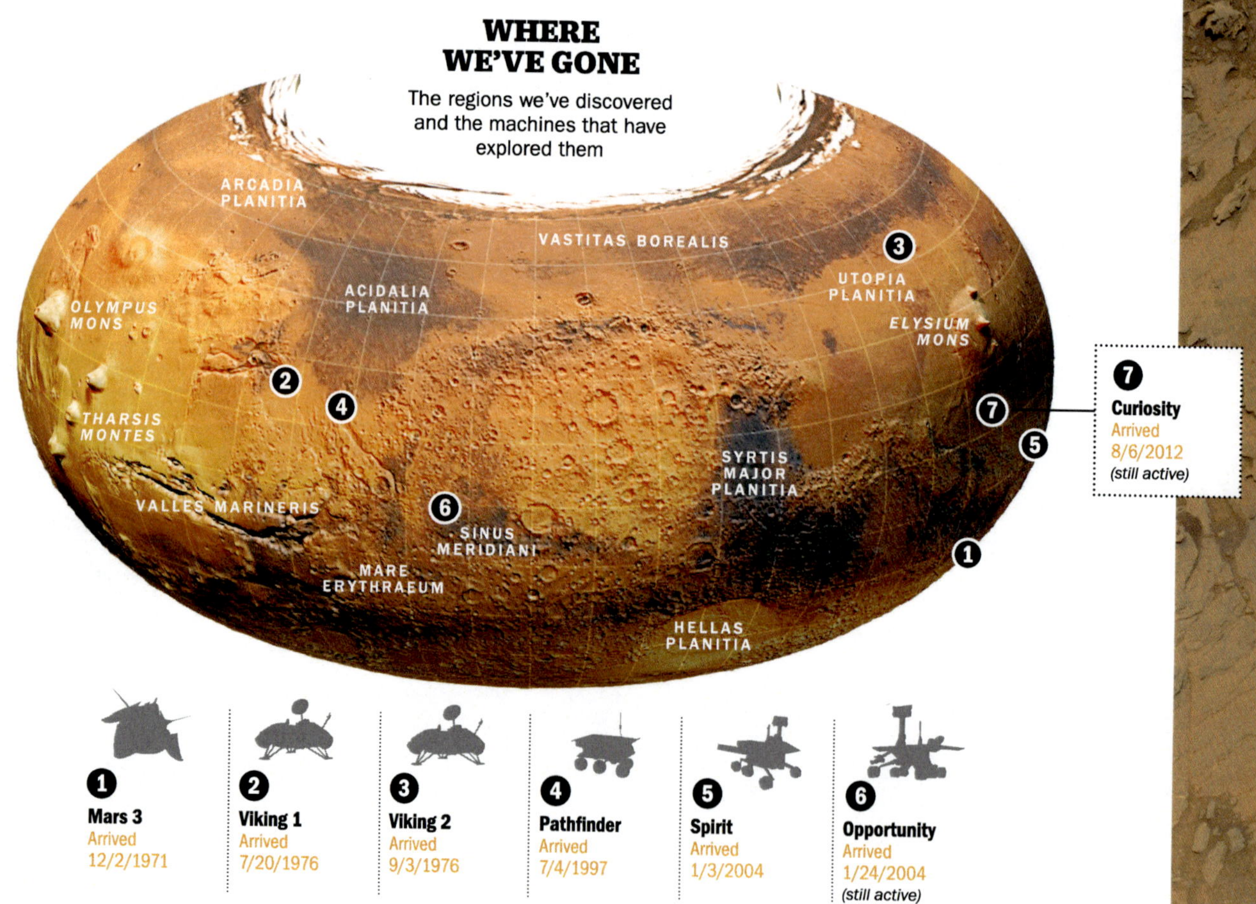

KORABL 4

U.S.S.R.; probe

Launched 10/10/1960

This is believed to be the Soviet Union's first attempt at a planetary probe, although some scientists involved with the Soviet space program claimed to have no knowledge of this mission. Korabl 4 reached an altitude of about 74 miles.

MARINER 4

U.S.; flyby

Launched 11/28/1964

Mariner 4 made history when it took the first photos of another planet from space. It delivered close-up images of the planet's surface when it made its closest approach at about 6,118 miles from the Martian surface on July 15, 1965.

1960s

1970s

MARINER 3

U.S.; flyby

Launched 11/5/1964

Designed to execute the first Mars flyby, Mariner 3 was supposed to capture photos of the planet's surface. But after a protective shield failed to eject upon passing through the atmosphere, the spacecraft never reached Mars.

MARINER 6

U.S.; flyby

Launched 2/25/1969

Mariner 6 took 24 near-encounter photos of the Martian surface, showing craters and dark features once thought to be canals.

MARINER 7

U.S.; flyby

Launched 3/27/1969

The orbiter, identical to Mariner 6, photographed new areas found during Mariner 6's flyby, such as Mars's south pole.

MARS 2

U.S.S.R; orbiter/lander

Launched 5/19/1971

The lander crashed after being released from its orbiter in November 1971, making it the first human artifact on Mars. The lander's braking rockets failed, but the orbiter returned useful data until March 1972.

MARINER 9

U.S.; orbiter

Launched 5/30/1971

Mariner 9 became the first U.S. spacecraft to orbit another planet when it reached Mars in November 1971. It produced the first detailed views of the Martian moons, Phobos and Deimos, and a canyon system larger than the Grand Canyon. Mariner 9 also mapped 85% of the planet's surface.

MARS 5

U.S.S.R.; orbiter

Launched 7/25/1973

Mars 5 captured about 60 images of the region below Mars's Valles Marineris after it entered orbit in February 1974.

MARS 6

U.S.S.R.; flyby/lander

Launched 8/5/1973

Mars 6 relayed data before losing contact as it neared the planet's surface, but much of it was unreadable.

PHOBOS 1

U.S.S.R.; orbiter/lander

Launched 7/7/1988

Phobos 1 was meant to study the surface and atmosphere of Mars and the interplanetary environment. It functioned properly until a command error caused it to lose its lock on the sun.

1980s

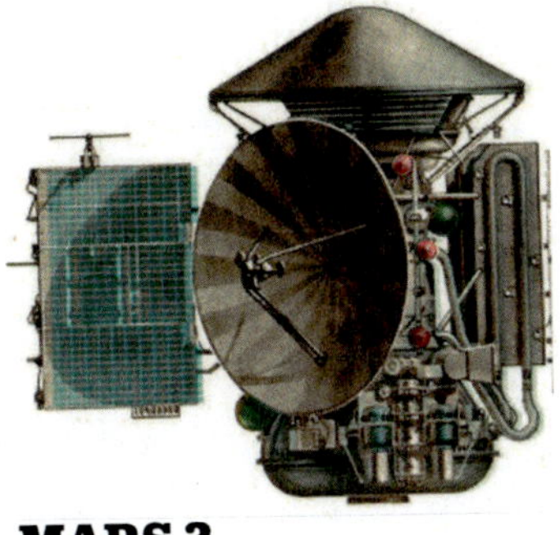

MARS 3

U.S.S.R; orbiter/lander

Launched 5/28/1971

The Mars 3 reached Mars on Dec. 2, 1971, and its lander became the first spacecraft to successfully touch down on the planet. The lander failed after relaying 20 seconds of video.

VIKING 1

U.S.; orbiter/lander

Launched 8/20/1975

This was the first U.S. spacecraft to safely land on the surface of Mars. In addition to returning photos from the surface, it conducted experiments that led to new discoveries about the chemical activity in Martian soil.

VIKING 2

U.S.; orbiter/lander

Launched 9/9/1975

The Viking 2, along with its Viking 1 sibling, found carbon, oxygen, nitrogen and other elements that support life on Earth during its time on Mars.

MARS OBSERVER

U.S.; orbiter

Launched 9/25/1992

Seventeen years after the Viking missions, NASA launched Mars Observer to learn about the geology, geophysics and climate of Mars. Contact was lost with the spacecraft on Aug. 22, 1993, just before Observer would have entered orbit around Mars.

NOZOMI

Japan; orbiter

Launched 7/4/1998

Japan's first Mars explorer was sent to research the planet's upper atmosphere, but the spacecraft ended operations in 2003 after a problem with its propulsion system prevented it from entering Mars's orbit.

MARS ODYSSEY

U.S.; orbiter/lander

Launched 4/7/2001

Odyssey has been circling the Red Planet since it reached its destination in October 2001. It's gathered data on the chemical elements and minerals that make up the planet's surface.

1990s **2000s**

MARS GLOBAL SURVEYOR

U.S.; orbiter

Launched 11/7/1996

Surveyor's key findings include an ancient water delta and active water features in canyon walls on Mars.

MARS PATHFINDER/ SOJOURNER

U.S.; lander/rover

Launched 12/4/1996

The Pathfinder mission delivered an unprecedented amount of data and far outlasted its expected life span. Findings from the lander and rover suggest that Mars was once a warm and wet planet.

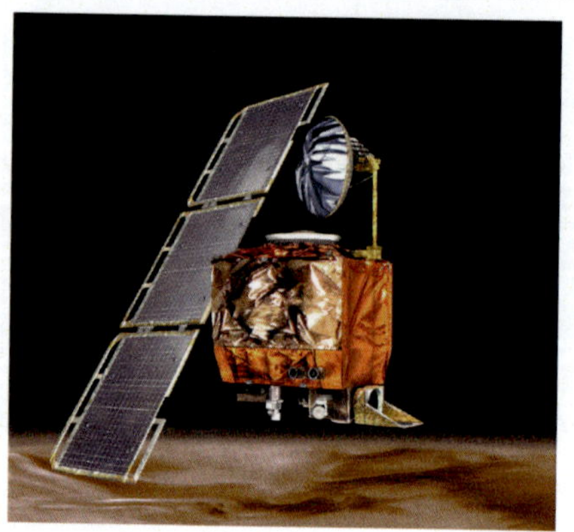

MARS CLIMATE

U.S.; orbiter

Launched 12/11/1998

Confusion over whether to use English or metric units destroyed this orbiter.

MARS EXPRESS/ BEAGLE 2

Europe; orbiter/lander

Launched 6/2/2003

Marking the European Space Agency's first voyage to another planet, Mars Express Orbiter has relayed thousands of 3-D views of the surface, from volcanoes to polar ice caps. Express entered orbit in December 2003 but lost contact with its lander, Beagle 2. The lander was found in 2015.

MARS EXPLORATION ROVER–SPIRIT

U.S.; rover

Launched 6/10/2003

Spirit found evidence suggesting that Mars was once much wetter than it is today. It has been stuck in soft soil since 2009, and NASA ended its mission in 2011.

PHOBOS-GRUNT/YINGHUO-1

Russia; lander

Launched 11/8/2011

The Phobos-Grunt spacecraft was intended to retrieve a sample from the Martian moon Phobos for observation on Earth, but it never exited Earth's orbit. It splashed into the Pacific Ocean in 2012.

MARS RECONNAISSANCE

U.S.; orbiter

Launched 8/12/2005

This orbiter, meant to study water on Mars, has one of the largest cameras ever flown on a planetary mission.

MARS SCIENCE LABORATORY

U.S; rover

Launched 11/26/2011

Better known as Curiosity, this rover is tasked with determining whether Mars was ever capable of supporting microbial life. It found that Mars once had more oxygen in its atmosphere, indicating a somewhat Earth-like past.

MARS ATMOSPHERE AND VOLATILE EVOLUTION (MAVEN)

U.S.; orbiter

Launched 11/18/2013

The craft is studying Mars's upper atmosphere with the goal of learning more about the planet's past climate, liquid water and habitability. It was the second mission selected for NASA's Mars Scout program.

2010s

MARS EXPLORATION ROVER– OPPORTUNITY

U.S.; rover

Launched 7/7/2003

This rover has continued to make crucial findings since it landed in 2004, including those that suggest the presence of saltwater.

MARS ORBITER MISSION

India; orbiter

Launched 11/5/2013

Representing India's first interplanetary mission, MOM was designed to observe the surface of Mars from orbit.

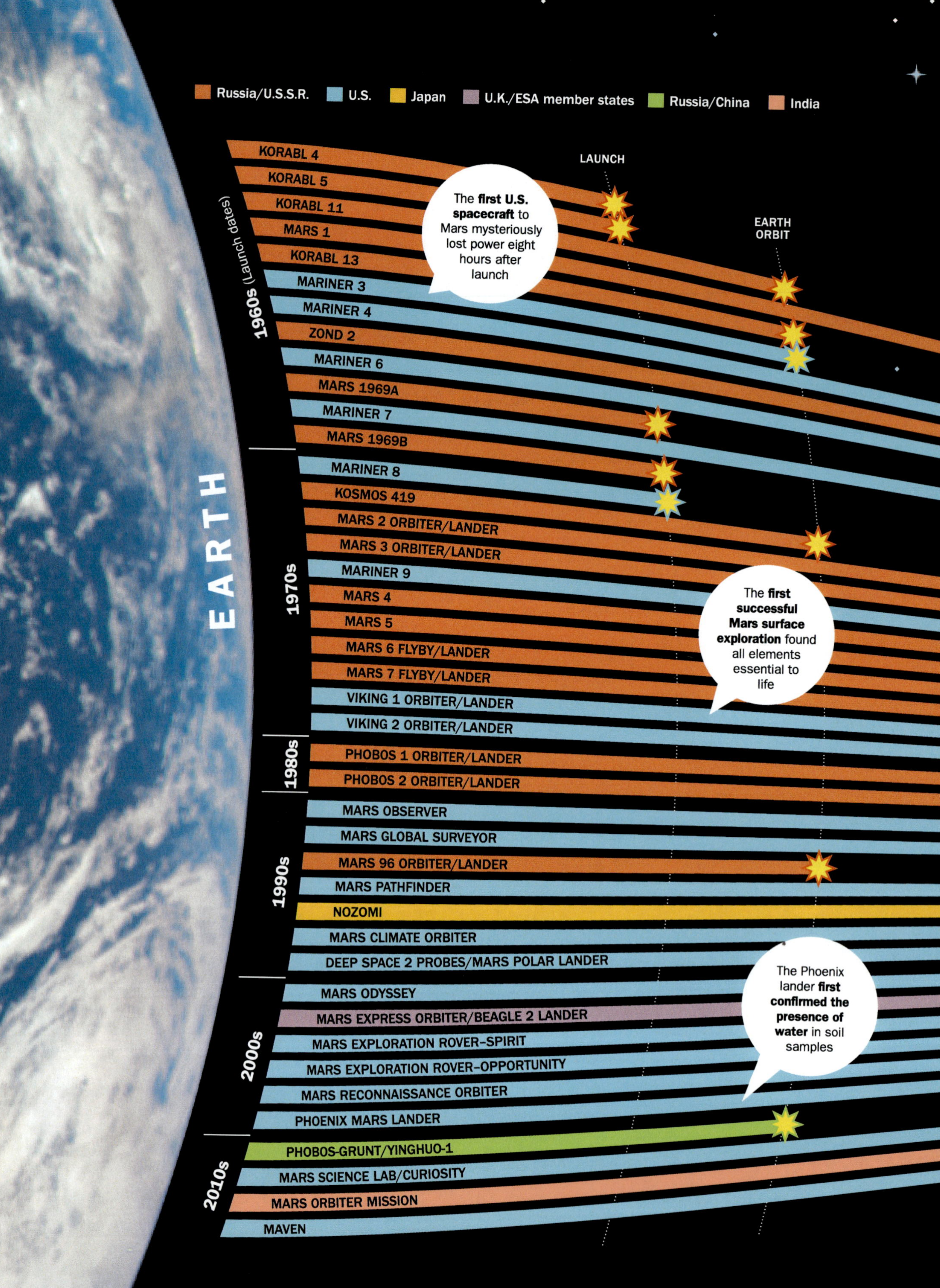

Russia/U.S.S.R.
U.S.
Japan
U.K./ESA member states
Russia/China
India

LAUNCH
EARTH ORBIT

1960s (Launch dates)
KORABL 4
KORABL 5
KORABL 11
MARS 1
KORABL 13
MARINER 3
MARINER 4
ZOND 2
MARINER 6
MARS 1969A
MARINER 7
MARS 1969B

The first U.S. spacecraft to Mars mysteriously lost power eight hours after launch

EARTH

1970s
MARINER 8
KOSMOS 419
MARS 2 ORBITER/LANDER
MARS 3 ORBITER/LANDER
MARINER 9
MARS 4
MARS 5
MARS 6 FLYBY/LANDER
MARS 7 FLYBY/LANDER
VIKING 1 ORBITER/LANDER
VIKING 2 ORBITER/LANDER

The first successful Mars surface exploration found all elements essential to life

1980s
PHOBOS 1 ORBITER/LANDER
PHOBOS 2 ORBITER/LANDER

1990s
MARS OBSERVER
MARS GLOBAL SURVEYOR
MARS 96 ORBITER/LANDER
MARS PATHFINDER
NOZOMI
MARS CLIMATE ORBITER
DEEP SPACE 2 PROBES/MARS POLAR LANDER

2000s
MARS ODYSSEY
MARS EXPRESS ORBITER/BEAGLE 2 LANDER
MARS EXPLORATION ROVER–SPIRIT
MARS EXPLORATION ROVER–OPPORTUNITY
MARS RECONNAISSANCE ORBITER
PHOENIX MARS LANDER

The Phoenix lander first confirmed the presence of water in soil samples

2010s
PHOBOS-GRUNT/YINGHUO-1
MARS SCIENCE LAB/CURIOSITY
MARS ORBITER MISSION
MAVEN

MISSION POSSIBLE

About half of all Mars missions have succeeded. Here's a complete history, up to the ExoMars Trace Gas Orbiter

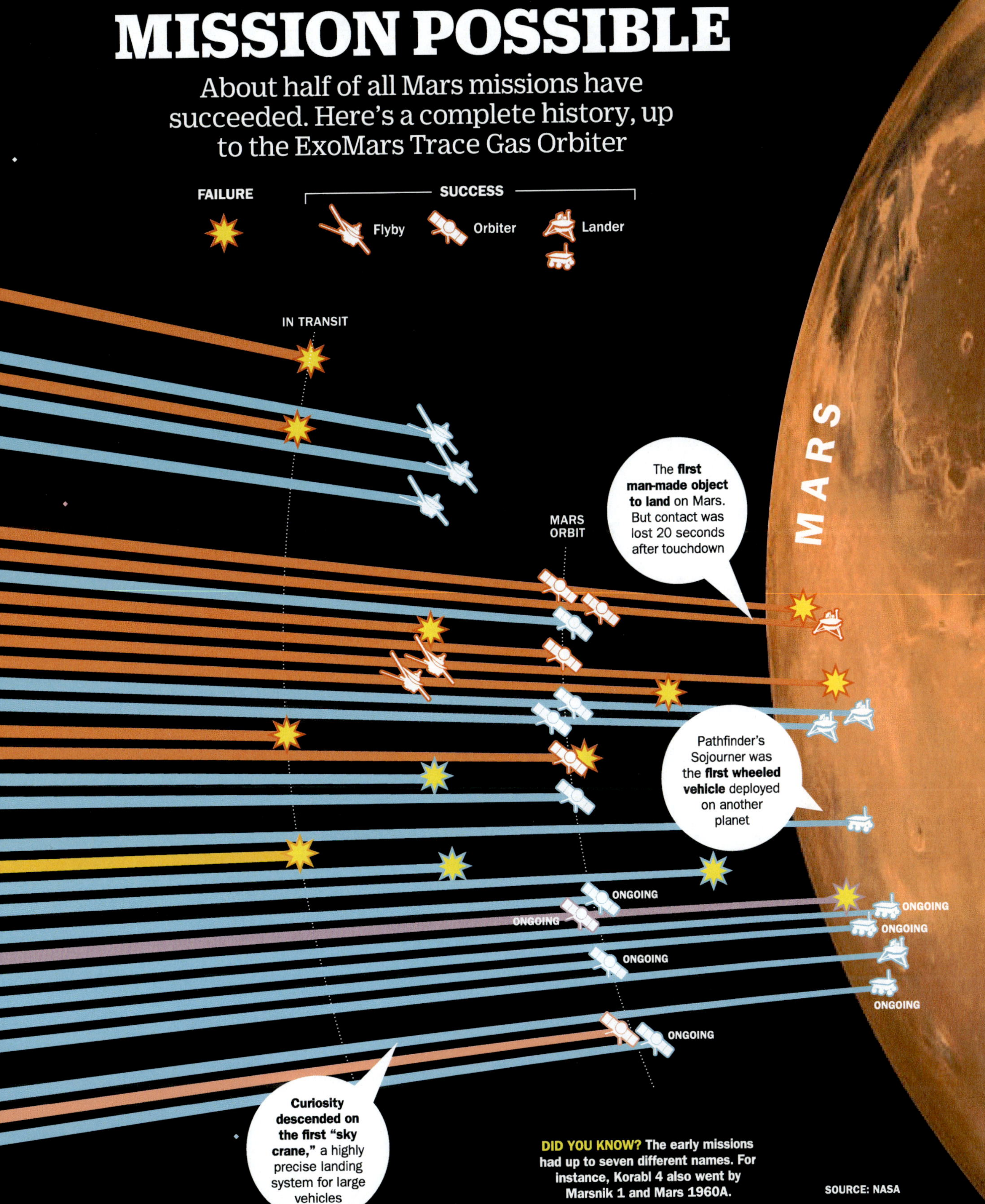

NEIGHBORS IN THE SKY

The love and attention that Mars gets from earthlings makes sense: the planet's proximity and biological potential render it irresistible. Yet there's a marble bag of colorful globes orbiting the sun, and NASA has sent spacecraft to every planet, plus Pluto. Each of these celestial bodies has taught scientists about the origin of the solar system and, by extension, the universe. Setting aside Mars and Earth for a moment, here's a little lowdown—along with magnificent images taken by NASA probes—on our solar system's most dazzling orbs

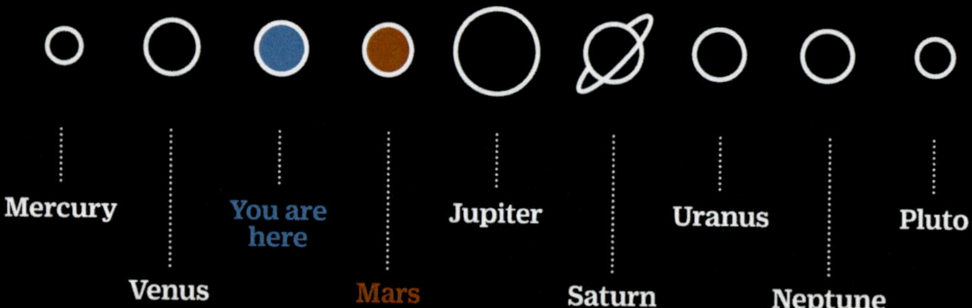

MERCURY

Distance from Earth:
48 million miles

How long to get there:
5 months

Proximity to the sun has its price: most of Mercury's outer layers have been blowtorched away, so that its core—mostly iron—makes up about 75% of its diameter.

Distances from Earth are averages. Transit times are based on NASA probes that have flown direct routes.

VENUS

Distance from Earth:
26 million miles

How long to get there:
4 months

A runaway greenhouse
effect means the
surface temperature on
Venus tops 860°F. Its
mostly carbon dioxide
atmosphere is 90 times
as thick as Earth's.

JUPITER

Distance from Earth:
483 million miles

How long to get there:
18 months

Giant Jupiter is one of the least dense planets, due to its gaseous makeup. Its powerful gravity protects the inner solar system by deflecting or attracting comets and asteroids.

SATURN

Distance from Earth:
874 million miles

How long to get there:
3 years, 2 months

Saturn's rings measure
a total of 175,000 miles
from inner to outer edge—
three quarters of the way
from Earth to the moon.
But the rings are only
30 feet thick at points.

URANUS

Distance from Earth:
1.79 billion miles

How long to get there:
8 years, 5 months

A gas giant, Uranus was the victim of one or more space collisions billions of years ago, knocking the planet on its side. A fragmentary ring encircles the equator.

NEPTUNE

Distance from Earth:
2.8 billion miles

How long to get there:
12 years

Neptune needs 165 years
to orbit the sun, but it
rotates much faster—
every 15 hours. Oh, and
bring earmuffs: at −360°,
it's one of the coldest
spots in the solar system.

PLUTO

Distance from Earth:
3.7 billion miles

How long to get there:
9 years, 6 months

In 2015, the New Horizons probe flew by Pluto at a distance of 7,750 miles. Demoted to a "dwarf planet" in 2006, Pluto is a complex world that may have subsurface water.

TWO MOONS

There's a solar system within the solar system, with at least 173 moons circling the planets and Pluto. (Mercury and Venus are moonless.) Some, like the two around Mars, are big bits of rubble. Others would be certified as planets if they orbited the sun themselves. And a few, such as the two here, are of particular intrigue, suggesting the possibility of life

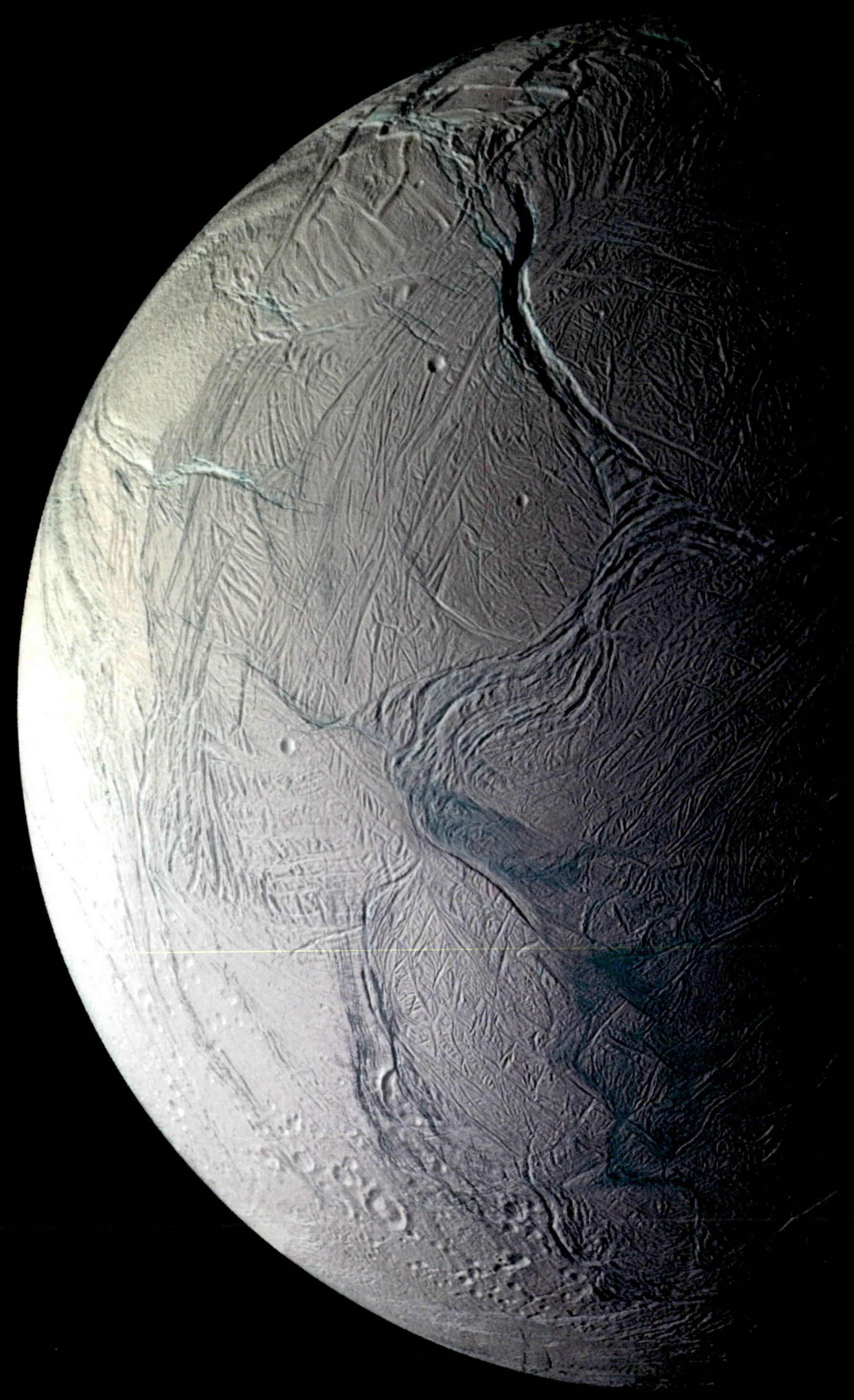

EUROPA

Distance from Earth:
483 million miles

How long to get there:
18 months

A moon of Jupiter, Europa is a prime candidate for life, thanks to the warm, salty ocean beneath its icy crust. NASA has plans to send a probe to Europa as soon as 2022.

ENCELADUS

Distance from Earth:
874 million miles

How long to get there:
3 years, 2 months

A moon of Saturn, Enceladus also has liquid water beneath its surface. Active in a geological sense, its frost geysers feed one of Saturn's rings.

CRASH LANDING
The sharp rim of this impact crater, formed when a meteorite hit Mars, suggests that it could be relatively recent. The steep inner slopes are carved by gullies, which could be a sign that water is seeping down these slopes.

CHAPTER TWO
THE PLAN
MANNED MISSIONS
TO MARS ARE
WELL UNDER
CONSTRUCTION

NASA'S NEXT STEPS

With a new crew vehicle and rocket boosters being built, NASA is getting its strategy off the ground

By Jeffrey Kluger

You don't want to be anywhere near the Cape Canaveral launchpad when NASA at last sends astronauts to Mars. You don't want to be near the 150-decibel roar the rocket will produce—a sound loud enough to rupture an eardrum. You don't want to be anywhere near the seismic vibrations that will rattle windows and knock down ceiling panels miles away and could be detected by a seismograph as far away as New York. You definitely don't want to be anywhere near the millions of pounds of fuel that, if something goes wrong, will explode with the force of half a kiloton of TNT, producing a fireball that would last up to 40 seconds and reach a temperature of 2,500°F.

And that might be the least of it. That level of sound and heat and power and menace was what was produced when the Saturn V booster launched men to the moon in the 1960s and '70s. That rocket loomed more than 36 stories high, weighed 6 million pounds fully fueled

Workers at Kennedy Space Center prepare to send Orion to Lockheed for more tests.

The second and final qualification motor test for the Space Launch System's booster was held on June 28, 2016, at Orbital ATK's test facility in Promontory, Utah.

and produced 7.5 million pounds of thrust. A Mars rocket, by all estimates, will have to be bigger and louder and more violent still.

That rocket, in turn, is only part of the puzzle. Also essential is a spacecraft that can be home to the crew as they make their trip. The first American astronauts had the little Mercury and Gemini spacecraft. The lunar astronauts had the Apollo spacecraft, the sturdy, conical orbiter that ferried crews to and from the moon nine times in relative comfort and reasonable safety—and brought them all home healthy and well to splashdowns in the ocean. A Mars ship, built for a much longer trip, would have to do the Apollo pod many times better.

But if NASA has proved one thing over its nearly six decades, it's that, given the funding and the backing, it is very, very good at building space machines—and the work on those Mars vehicles is now well under way. The new booster is prosaically known as the Space Launch System, or SLS; the new crew vehicle is more ambitiously named Orion, after the most heroic constellation in the night sky.

Whatever their names, these ships are the real deal. All over the country—in factories and assembly plants in Louisiana, Virginia, California, Mississippi, Alabama and elsewhere—metal is being cut and systems are being built for what will easily be the most ambitious and powerful flying machines ever invented. Their trips to Mars may still be 15 or more years away, but their preliminary launches—some of which will take humans deeper into space than we've ever gone before—are coming up much sooner. More than 40 years after human beings last went to the moon, the next generation of deep-space vehicles is at last slowly making its way to the launchpad.

BUILDING THE MACHINES NECESSARY FOR astronauts to make the eight-month trip to Mars—which is the minimum time the journey takes when the planets make one of their periodic close approaches—may be harder in almost every respect than building the machines for a mere three-day trip to the moon. But we've got one edge now that we didn't have then: we al-

ready have a lot of the parts in hand. The Saturn V was basically built from the ground up—made of entirely new components invented for an entirely new purpose. The SLS will be able to rely on a lot of components already on the shelf, and it will grow over the decade from a test version to the fully outfitted Mars model.

The first functioning SLS, which could launch as early as 2018, is known inside NASA as the Block 1, and it's the smallest of the SLS family—but "small" is decidedly relative. The runt of the SLS litter will stand 322 feet, just 40 feet shorter than the Saturn V was. Like the Saturn V, it's a traditional, expendable rocket—beginning its mission with an upright launch and ending it with an ocean splashdown, unlike the reusable space shuttles, which launched upright but glided to their landings like an airplane. All the same, the SLS's engineering DNA is sunk deep in the shuttle's.

The rocket will produce a whopping 8.8 million pounds of thrust at liftoff—comfortably more than the Saturn V's 7.5 million pounds—and much of that muscle will come courtesy of two solid rocket boosters attached to either side of the main body of the rocket, very much like the ones that were positioned on the shuttle. The SLS's solids will be about 10% more powerful than the shuttle's were, and each will produce 3.6 million pounds of thrust.

Solid rockets make some people nervous, and not without reason: flames leaking through a breach in a solid ignited the fuel tank of the shuttle Challenger, destroying the spacecraft and claiming the lives of its crew during a launch in 1986. What's more, solids, which burn a dense, rubbery fuel, can't be throttled up and down the way a rocket that uses fuel like liquid hydrogen and liquid oxygen can. They're either on or off, operating at full power or not at all.

The simplicity and affordability of solids is undisputed, however. And with the terrible exception of Challenger, they've had an almost spotless record for 30 years. "We need a heavy-lift launch vehicle and solids to get us part of the way there," says Jason Crusan, NASA's director of Advanced Exploration Systems. "They're a very established technology."

NASA will get the rest of the way to liftoff with more off-the-shelf shuttle technology: a cluster of four RS-25 liquid-fueled engines at the bottom of the rocket itself (the shuttle used three), making six engines in all to get the SLS off the ground. The liquids, too, will be souped up to wring out more power than the original models had. Sixteen of the original RS-25s are left over from the shuttle days, so they would only have to be modified and performance-rated to be good for the first four SLS flights.

Even with all that engine oomph, the Block 1 SLS will not be enough for a Mars mission. The 2018 test flight, which will use a Block 1, will carry an Orion spacecraft to an altitude of only 3,000 miles. That's just 1.3% of the 230,000-mile distance to the moon, and never mind the percentage of the 45 or so million miles that separate Earth and Mars even when they're making one of their close approaches.

Still, the flight will be a good test of both the booster and the spacecraft, and any SLS launch is a good launch. Flat NASA budgets have led to repeated delays in the SLS program, with talk of a first flight by 2015 giving way to promises of 2016 and then 2017. But NASA's 2016 budget of $19.3 billion is a more than $1 billion bump over the previous year's, and it is no secret how a lot of that money was intended to be spent.

"NASA is firmly on a journey to Mars," administrator Charles Bolden announced when the budget request was unveiled. "Make no mistake, this journey will help define and guide our generation."

That kind of official backing has given the space agency the confidence to proceed with the development of its later, bigger SLS models even as it's still working to launch the test version. The first of the upgrades will be the Block 1B, equipped with a second stage that will be able to carry heavier payloads and send astronauts to deep space, if not quite to Mars yet. The Block 1B will also top out, conveniently, at 364 feet—or exactly one foot taller than the Saturn V. The first crewed flight of the Block 1B will come sometime around 2023 and will be a multi-day flight go-ing beyond the moon and coming back home. A Block 2, with even bigger solid rockets, will carry astronauts to Mars in the 2030s.

If all of that seems complex and slow and incremental, it's because it is—but complex and slow and incremental works. The Saturn V was not, despite the name, the fifth model of its kind; it was preceded by Saturn 1 and a Saturn 1B. That kind of stepwise progress was necessary to make the final rocket the reliable machine it was.

A ROCKET, HOWEVER RELIABLE IT MAY BE, gets you only so far if you don't have a vehicle on top that can carry a crew. Orion, like the SLS, is making slow but consistent progress toward flightworthiness. The spacecraft bears more than a passing resemblance to the conical design of its Apollo ancestor, though it includes much more sophisticated materials, electronics and navigation systems. It's also bigger, with a habitable volume of 692 cubic feet as opposed to the Apollo's more cramped 218.

More volume means bigger crews, but NASA doesn't plan to take advantage of every extra foot. Orion can be configured to carry six people—compared with Apollo's three—for relatively brief missions. For a Mars mission, the crew manifest will probably be just four. That's enough people to get the job done, while allowing more room for comfort. In theory, a three-person crew could do the job too, risking fewer lives and making the cockpit even roomier, but international politics makes that tricky.

The U.S. has collaborated with 14 other nations to build, fly and maintain the International Space Station, and the hope is that much of that same coalition will collaborate for Mars, spreading out the work and the cost. That, though, means competition for a seat in the spacecraft. "We want to make sure there are enough crew members to help represent our international partners," says Crusan. "We're starting with the ISS partners we have on hand, and those discussions are under way."

Orion, unlike the ISS, has had one test flight. In 2014 it was launched by a Delta

IN THEORY, A THREE-PERSON CREW COULD DO THE JOB. POLITICS MAKES THAT TRICKY.

rocket—built by the private company United Launch Alliance—on a quick, high-altitude orbit that allowed it to test the high-speed reentry a spacecraft returning from Mars will have to navigate. A very fast reentry also means a very hot reentry, with the ship's heat shield subjected to temperatures as high as 4,000°. Orion took it all well, ending its mission with a flawless splashdown in the Pacific and boosting confidence in the program as a whole.

NASA's human-exploration initiative has been adrift for a long time, with projects getting proposed and canceled, proposed and canceled again and again over recent decades. But Orion and SLS appear to have brighter prospects. Too much hardware has been built and too much capital has been spent—both real and political—for the projects to be canceled without officials in Washington facing a firestorm of criticism. But even if the shiny new machines do get built and launched, there are still a lot of pieces missing.

Above: Test dummies are seated in an Orion test capsule before being dropped into the Hydro Impact Basin at NASA's Langley Research Center in Hampton, Va. Right: The Orion Ground Test Article completes its first swing splash test at Langley. Below: A test version of the Orion spacecraft takes a dive into the Hydro Impact Basin.

An eight-month journey to Mars is far too long for four astronauts to spend all their time inside a spacecraft no larger than the interior of an SUV. That means a habitation module— similar to the school-bus-size segments that make up the International Space Station—will have to go along for the ride. There's a landing and liftoff vehicle similar to the Apollo lunar modules so that the astronauts can get down to the surface of Mars and later depart. And then there is the habitat in which the astronauts will have to live while they're on the ground and a rover so they can get from place to place.

"You can stay a few weeks and take advantage of the same Earth-Mars conjunction," says Crusan. "Or you can stay longer, but then you're there for a year until the planets line up again. The longer the stay, the more food and supplies you need and the bigger the habitat has to be."

None of these challenges come as a surprise to NASA, and none will be overcome quickly. It's fitting, however, that the design and construction work begins where the mission itself will begin: with a rocket and a spacecraft destined for a launchpad on the east coast of Florida. Orion and SLS may not be all we need for a journey to Mars, but the steady progress of their development is a powerful sign that we're committed to making the trip.

THE WHITE HOUSE SCIENCE CZAR GETS GALACTIC

John Holdren
on traveling to
Mars, redirecting
asteroids, Obama's
legacy and more

By Jeffrey Kluger

I t's not easy being John Holdren. As director of the White House Office of Science and Technology Policy, he has the unenviable job of explaining and promoting complex policy to a public that too often rejects what ought to be basic scientific truths. But one thing most Americans still agree on is space: we love it, we're good at it, and we can't get enough of it.

Since 2011, the American space program has been at a crossroads, with the shuttle in mothballs and the International Space Station (ISS) still flying, but with no means to send astronauts there without buying seats on a Russian Soyuz rocket—which go for more than $70 million each. And those challenges don't even take into consideration NASA's long-term goal of getting astronauts to Mars by the 2030s. With a change in presidential administration nearing, TIME sat down with Holdren earlier this year to discuss these and other issues.

The NASA budget, at less than 1% of federal spending, is nowhere near as robust as it was in the Apollo days. Can we really get to Mars—even by the 2030s?
Getting to Mars with humans is a big challenge. It's going to be expensive. I believe it will be done jointly—it will be an international project. I don't think we will have a race to Mars with the United States, Russia, the European Union, Japan and China all competing. I think we are going to go together, and that is one of the reasons it will become affordable.

Is there any preliminary work or planning being done for that kind of international mission?
Of course there are preliminary conversations going on. We are not at a moment of great relations with Russia, but we still collaborate with Russia on the ISS, which is a great example of international cooperation in space. Roughly $100 billion has gone into that operation. It

really is an unbelievable project when you look at what is almost a small town in the sky. Mars will be a bigger project, in all likelihood, and I expect again that all of the major players are going to come to the view that sending humans to Mars simply has to be a joint venture.

The wild card that exists today and that didn't exist in the Apollo era is the private sector. Could individuals and private companies such as Elon Musk and SpaceX overtake government space programs?
We are going to see an increasing private-sector role in space. We have brought in new models, working with the private sector to get cargo and humans to the ISS. The initial idea was to give the private sector a larger role in relatively straightforward missions like going to LEO [low-Earth orbit] while letting NASA focus its capabilities on its most demanding missions. But the private sector is proving so energetic and ambitious that I have to assume that [it] will be looking to play a bigger role in the more demanding missions.

President Obama has articulated a nearer-term goal of an asteroid redirect mission—the idea being to capture a small asteroid, steer it to the vicinity of the moon and then send astronauts to land on it. A lot of people have asked what the purpose of such an improbable mission is.
I actually think there is much less skepticism out there than earlier. I think it makes sense from the standpoint of developing techno-logical capabilities. It also makes sense from the standpoint of developing a better under-standing of the asteroids that come closest to the Earth and the possibilities for deflecting one should it come up on a collision course.

Also, a lot of commercial companies are interested in learning what kinds of minerals asteroids contain and possibly developing a business in mining them for materials that could be used in space, saving us the trouble of lifting them out of the Earth's gravitational field. There are even companies that believe there are going to be such valuable minerals that it would be worth bringing them back to Earth.

The moon has been taken off the table as a goal for manned spaceflight, but one of the challenges of a Mars mission is figuring out how to build a base camp and live off the land for months at a time. Why doesn't it make sense to practice that first on a world that is only three days away?
The short answer is, fundamentally, money. Our estimate of what it would cost to develop a base on the moon is $60 billion or more. No one looking at the U.S. budget can figure out where the money to do that would come from. Basically, by operating in the vicinity of the moon as we would in the case of an as-teroid redirect mission, we can develop and demonstrate the capacities we would need for long-term habitation without paying the enormous price of setting up operations on the moon.

How confident are you that the next president will continue President Obama's space policies, particularly concern-ing the construction of a new heavy-lift rocket and crew vehicle for deep-space exploration?
There are certain fundamentals that every-one who looks at the challenges of space exploration [recognizes]: a heavy-lift rocket is one of them; a crew capsule is another. Every president who comes into office and looks at the space program also comes to understand what the program means for U.S. technological leadership and what it means as inspiration for young people. We still find that getting kids inspired about science and math and engineering is one of the great features of the space program. It seems to do it more effectively than anything else.

What will President Obama's space legacy ultimately be?
First of all, restructuring the space program and coming to an agreement with Congress so that we can meet congressional priorities and the president's priorities. We have put the [human exploration] program on a sustain-able course without gutting space science, as-tronomy, cosmology and the space telescopes, which are all immensely important.

MARS ON EARTH

Human psychology may be the biggest wild card on a mission to Mars. In an ongoing NASA program, six acting astronauts are spending up to a year in isolation on a simulated Mars base in Hawaii. It's not the Red Planet, but it's close

Regular geological expeditions keep the earthly astronauts busy. Spacesuits are required for all trips outside, just as they will be on Mars.

The six-person crew lives in a 13,000-cubic-foot habitat with a loft, separate sleeping quarters, a bathroom with a shower, and exercise and work spaces.

It takes a lot of food to feed six people for a year. Shelf-stable products make up much of the meals, but the crews also grow crops, like spinach and broccoli, in the habitat.

A YEAR IN SPACE

A historic mission began
to show how best to care
for a human body—and
mind—when it spends
too long defying gravity

By Courtney Mifsud

Astronaut Scott Kelly, with cosmonauts Mikhail Kornienko and Sergey Volkov, returns to Earth near Zhezkazgan, Kazakhstan, on March 2, 2016.

What's with all the overcoats?" Scott Kelly asked the search and rescue team that helped him from the Soyuz TMA-18M capsule. "This feels great." It was March 1, 2016, and Kelly, the NASA astronaut and Expedition 46 commander, had just touched down near the town of Zhezkazgan in still wintry Kazakhstan (hence the coats). Kelly, along with Russian cosmonaut Mikhail "Misha" Kornienko, had just completed an unprecedented 340-day stay aboard the International Space Station.

Kelly's landmark mission took steps toward assessing and addressing some of the physical limitations that stand in the way of further space exploration. Humans evolved for life on Earth, and when astronauts spend extended periods of time away from the comfort of its one-G environment, their bodies deteriorate. "NASA needs to understand the long-term impact of these missions in order to identify strategies to monitor health outcomes and

Kelly trains for his mission in a Soyuz simulator at the Gagarin Cosmonaut Training Center, just weeks before heading to space on March 27, 2015.

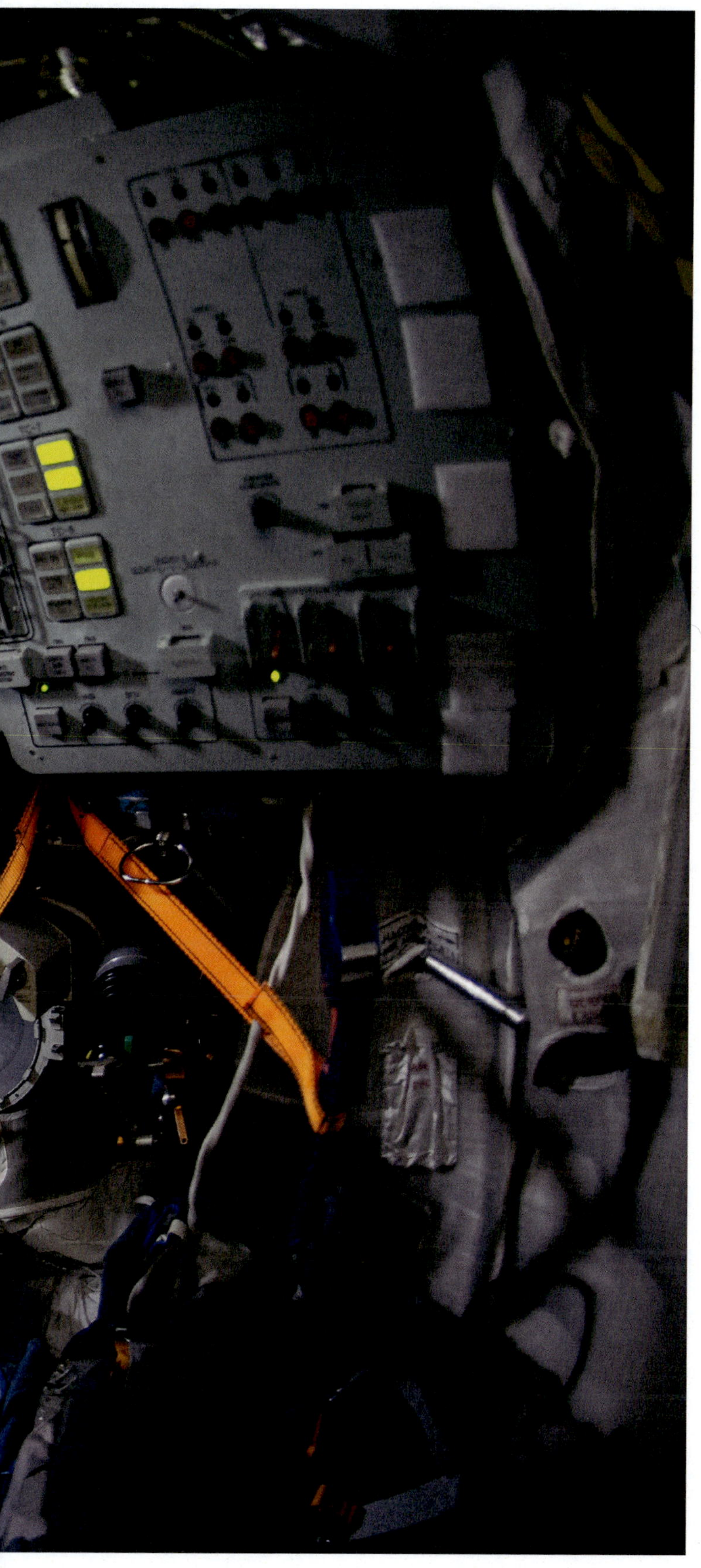

reduce health risks," says Brinda Rana, a professor at the University of California, San Diego School of Medicine who was part of the team that monitored the health effects of the one-year mission.

Gravity is crucial. Without its pull, bones begin to break down and eventually decalcify, with density dropping more than 1% per month. (In comparison, bone loss among the elderly is between 1% and 1.5% over an entire year.) Bones aren't the only issue: Human bodies are made up of 60% water, which makes them hugely unsuited to zero gravity. The liquid shifts upward, putting pressure on the chest, the head and especially the eyes. This pressure compresses the optic nerve and alters the shape of the eyeball, resulting in farsightedness. NASA doctors have put countermeasures in place. Compression cuffs worn on astronauts' legs help prevent the upward migration of blood, counteracting the vision problems. Bisphosphonates, a class of drugs used to treat osteoporosis, coupled with innovative exercises such as running while harnessed to a treadmill, can be used to help protect bone density. Yet there remain significant challenges to overcome before NASA sends humans to Mars.

That's where Kelly and Kornienko come in. They participated in 17 different physiological and psychological experiments on their mission, aimed at better understanding the effects of extended spaceflight. And in the case of Kelly, the universe had handed science the ideal control subject: his identical twin brother, Mark, a retired NASA astronaut with four ISS missions under his belt. Mark retired in 2011 and remained on Earth during his twin's year on the space station. As NASA collected samples from Scott, Mark provided the same. "I imagine that NASA is hoping to learn that space isn't a big deal, that it doesn't have that big of an impact on astronauts," said Mark during Scott's mission. "That would be great. But I think the reality is that it does, and so we're going to have to learn how to mitigate those effects."

The physiological experiments monitored changes in eyesight and the influence of certain nutrients. NASA watched any shifts closely, including identifying if spaceflight and the zero-G environment caused plaque buildup in Kelly's arteries. One experiment focused on the microbes that facilitate digestion. Scott and his crew received multiple shipments of fresh fruits and vegetables. They also germinated a batch of edible flowers, part of a larger plant system that will become crucial to a manned Mars mission. As Gioia Massa, a payload scientist at the technology company Orbitec, told TIME, "The farther and longer humans go away from Earth, the greater the need to be able to grow plants for food, atmosphere recycling and psychological benefits."

Six other twin studies focused on the environmental factors of space, notably the high levels of radiation exposure. Earth's magnetic field protects the planet from harsh cosmic radiation, but once astronauts leave the atmosphere, they're vulnerable. Over the course of his 11-month extraterrestrial spin, Scott Kelly was subjected to more than 10 times the amount of radiation as his brother on Earth. And a crew on a mission to Mars would endure far more. NASA operates with a restriction that its astronauts should not have their lifetime cancer risk heightened by more than three percentage points, meaning that if humans on Earth have a 21% chance of dying from cancer, NASA cannot allow that percentage to exceed 24% for their astronauts. Mark Kelly said he would take on twice that 3% added risk for the opportunity to go to Mars.

ANY STRESSFUL SITUATION SHORTENS LIFE expectancy, and spaceflight is no exception. But how do the stress levels created by spaceflight compare with the stress of an entirely Earthly event, such as the loss of a job? To help identify and predict the impact of stress levels, Susan Bailey of Colorado State University coordinated one of the major facets of the twin study, examining telomeres. Telomeres are stretches of DNA at the tips of chromosomes. They protect the chromosomes and essentially control aging. As we get older, our telomeres shorten and the chromosomes they're connected to begin to unravel. "A whole variety of life stresses have been as-

sociated with accelerated telomere loss as we age," Bailey says. "You can imagine strapping yourself to a rocket and living in space for a year is a very stressful event."

To facilitate the study, over the course of Scott Kelly's mission, he regularly drew his own blood in space, and his chromosomal samples were ferried back to Earth from the ISS on cargo ships to be compared with Mark's. The experiment continued after Scott landed and readapted to life on Earth. In particular, scientists are evaluating whether the potential telomere loss that Scott endured in space slows in the two years following his return home.

NASA is also concerned about the psychological effects of long-duration spaceflight. A year aboard the space station is incredibly confining, and decision-making ability, alertness and sound reasoning tend to wane during a long mission—especially between the halfway and three-quarter points. And consider that the ISS is close enough to Earth that emails can be sent and video chats can be had without lag time. Scott was never far from supporters at home rooting for him, and he frequently kept his followers up to date via social media.

Connecting with home via video chat might prove vastly more frustrating to a crew en route to Mars. On average, Mars is 140 million miles away from Earth; the ISS is just 249 miles away. Service would get spotty if astronauts tried to call mission control from Mars or even close to it, and real-time conversation would most likely be impossible. Keeping a journal helps astronauts psychologically (while also providing valuable feedback to mission control). Astronauts either type their journals or record audio files at least three times a week, giving them an outlet to express emotions.

Scott Kelly may have enjoyed the crisp air in Kazakhstan when that Soyuz hatch opened, but the rest of him wasn't feeling so great. "When I got out of the Soyuz . . . I didn't really look too bad," he told a NASA auditorium in his first major address, almost three months after his landing, "but that was only because I'm a very good actor. I think I should be nom-

> **"WHEN I GOT OUT, I DIDN'T LOOK TOO BAD, BUT THAT'S BECAUSE I'M A GOOD ACTOR."**

Kelly, after a delivery of fresh fruits and vegetables to the space station, posted this photo on Instagram.

inated for an Academy Award." When Kelly returned home, he said, he experienced flu-like symptoms and burning skin. "But that's why we do this," he said. "We need to learn these things if we're going to go to Mars."

Finding answers to key health questions will be necessary to make good on any promise to set foot on our planetary neighbor. Whether man's eventual mission to Mars is five years away or 500, Kelly and Kornienko's won't be the last one-year mission to help safely prepare for that journey. "We're looking at as many as 10," Doug Wheelock, the incoming director of NASA's Star City, Russia, office, told TIME. "We need a good mix of subjects, which means women and men, older crew members and younger ones, veterans and first-timers. There's a lot we have to learn."

HOW TO SNEEZE IN SPACE

A talk with NASA's top medical scientist, John B. Charles, delivers answers to some of the crucial health issues tied to life away from Earth

By Alexandra Sifferlin

hile the risks for health complications such as catching the flu or having a heart attack may be low during a trip to Mars—astronauts are in good shape to begin with, are quarantined for weeks before heading out and are unlikely to have strangers cough on them during cocktail parties along the way—NASA, of course, has to prepare for the possibility that someone gets hurt or sick. TIME spoke with John B. Charles, who has a Ph.D. in physiology and biophysics and has been with NASA since 1983.

What are some of the health complications NASA is prepared for in space?
There are concerns we have for astronauts, especially on long-duration spaceflights. Starting from the first moments of spaceflight, we have the change in the way the body senses the direction of down. The neurosensory organs—your organs of balance in your inner ear—are disrupted when you go through the launch and enter weightlessness. That can cause motion sickness. Your body is full of fluid, and when you are on Earth, that fluid tends to want to pool in the lower portions of your body. Spaceflight is sort of like being recumbent all the time without being upright. So there's a possibility that you will have a shift of body fluids into the upper part of the body, which leads to puffiness of the face. Astronauts often talk about their puffy faces and skinny legs.

You also have psychological effects. Being confined to a small vehicle with a small number of other people for long periods of time can be disruptive. Still, you get to see Earth and you get to be weightless. It's something really important and exciting and historic, and some people are willing to take these risks.

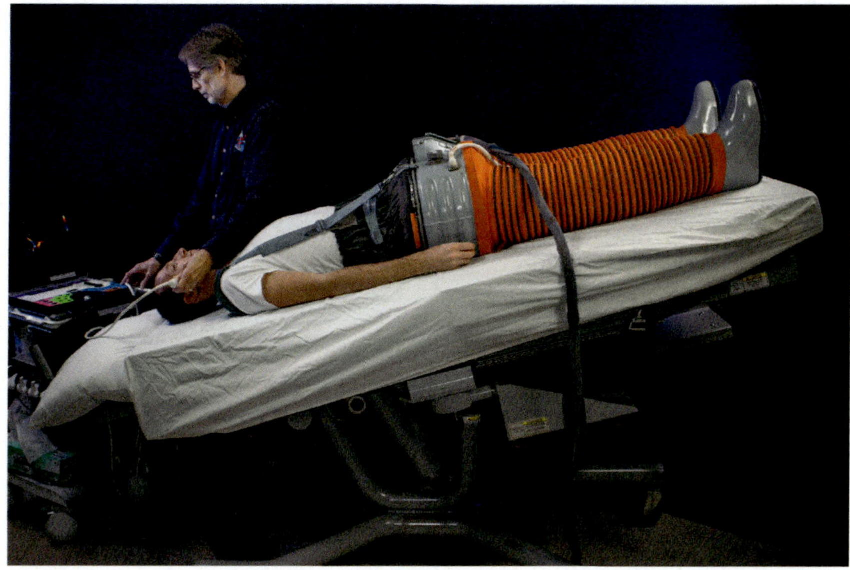

When people go to Mars, will there be a medical team with them?

They will have to *be* the medical team. They will have to be the engineering team, the geology team, the cooking team, the piloting team and the public-affairs team. Luckily, NASA is good at finding people who are good at multiple things. I like to joke that the Mars crew will have to include a neurosurgeon, a concert pianist and a gourmet chef. Clearly there will have to be medical capabilities for the crew members. They will have to be outfitted appropriately. One of the interesting problems that we are facing in our early planning for Mars missions is, What can we protect for? What can we provide the astronauts medical capabilities for, and when do we just say, "We can't help you"? There's not an ER on the ground. You have four people. Bad things are going to happen, and that's why the crew members are going to need to be sufficiently experienced and sufficiently motivated and sufficiently trained to know which is priority.

What if someone breaks an arm or a leg?

If they are on Mars outside the habitat, they are essentially inside an airbag. They are inside of a balloon suit. If the suit is like our current suit, it is going to be very stiff. If they break an arm or a leg, it's because something seriously bad happened, like maybe they got their arm or leg wedged inside a crevice or they fell off a cliff. The worst thing that happened to you probably isn't breaking a leg. But say you were in the habitat [not in your space suit] and the furniture was arranged inconveniently, and you're bouncing along and bump into a corner and it does break a limb. There will have to be splints on board, and [the crew members] will all have to have the basic medical skills to do things like splinting a broken bone and relocating or reinserting a dislocated finger or shoulder. They all get trained in these kinds of things in a very rudimentary style. If you splint [the broken bone], it will eventually heal.

What happens if you need to sneeze in a space suit?

People do sneeze in suits. When you sneeze in your suit, then you have your sneeze on the faceplate in front of you for the rest of the time you're on your space walk. If that kind of thing makes you queasy, then you probably shouldn't be in the space suit in the first place, because these things are inevitable. You may have had a substantial sneeze. You might have [the sneezed mucus] hanging between your nose and the visor for hours and hours at a time. Imagine trying to be heroic and saving the space station while this is going on. This just happens. You try not to sneeze, or you try to aim your sneezes down inside your suit so you don't mess up your visor, but astronauts do say it happens. When it does happen, you just tough it out and do the job.

What if your appendix bursts?

That shouldn't happen, because if you have an inflamed appendix, you shouldn't be assigned to the flight. There have been people who have advocated appendectomies before flight, but surgery is a risk as well. What if you take out a perfectly normal appendix and you make the person worse off by the surgery and then the person is disqualified as an astronaut? I have been told by doctors in the past working here at NASA that, number one, you select for healthy appendices or, two, you provide the appropriate antibiotics so that if someone does get an inflamed appendix they can address it medically with pharmacology instead of surgery.

In the last decade or so, there have been a lot of improvements in surgeries. There are capabilities now on Earth to use small incisions and inserted probes that can reach in and do the job, and NASA is investigating those things. So that might be part of the surgical capability on a Mars mission for cases where it really can be helpful and useful. But surgery is always going to be the last resort in a spaceflight.

How about heart attacks?

Don't hire a person who is going to have a heart attack. They are not always predictable, and things can change in flight. If it does happen, I know NASA, in the shuttle era, did fly an automatic defibrillator on at least a few of the missions because they were concerned about that. But it's never been cracked open or activated, and people don't have heart attacks in spaceflight. You can never say never, but that's the kind of thing you really just hope doesn't happen. If it does happen, then you treat the person medically: you provide all the sedatives, all the fluids, and let them stay in their sleep quarters and keep an eye on them. You're hoping that the heart muscle will recover on its own, as it often does, and then

they can resume their role as a useful person on the spaceflight.

Can you catch the flu in space?

If you brought the virus with you, then yes, you can. Prevention is always better than a cure, but there are problems with those kinds of things. You isolate the crew members before flight, but sometimes people come in who were not successfully isolated. Sometimes astronauts break isolation because they just have to see the kid's baseball game or the concert or piano recital. There's also the possibility that the spacecraft has contaminants including viruses and bacteria on it. The astronauts who go to Mars will not have been the first people on that vehicle. There's a chance that those crew members and the vehicles that bring up supplies bring contaminants. It's not impossible. Also there's the flora that live in our GI system and the things that live in our respiratory systems. They can get out and find niches in spaceships and grow and multiply. It will be important to keep the spacecraft scrubbed down and clean and make sure you have the appropriate air and water filtering and take your basic precautions like we all do on Earth.

Is it possible to get food poisoning from space food?

That can happen if we are not careful. There have been examples of food on the International Space Station spoiling. The cans or the packages were not treated correctly. Maybe they got overheated or dropped in shipment and were ruptured or cracked. Food packages have spoiled, and the astronauts don't eat them. They throw them away. You have to be careful of what you are eating. If a bag is loaded with gas or a can hisses when you open it, throw it away and don't eat it. If you do get food poisoning, there are medical ways to treat it involving hydration and leaving the person alone in the bathroom for a long period of time so they can do what Mother Nature requires. Clean up afterward, give them a few days to

> **"I AM VERY STRONGLY CONCERNED ABOUT PEOPLE WHO WANT TO FLY AND HAVE BABIES IN SPACE."**

recover and get their body fluids back, and let them move back into the normal work cycle.

What if someone gets pregnant in space? They should be able to make sure it doesn't happen. They are all normal, healthy men and women, and if circumstances conspire for someone to get pregnant, well . . . it's going to be one of those situations where prevention is better. I think personally that's an ethical question as well as a medical question. It's a high-radiation environment, a weightless environment, and there's a lack of appropriate nutrition. Assuming the fetus grows normally and delivery occurs, then you suddenly have another crew member to take care of. One who is very high-maintenance and who is developing and growing by the minute. When that baby comes back to Earth, the baby will be at a deficit in terms of bone strength since it will not have had gravity during the formative phases. There are better things to do than having babies in space. I am very strongly concerned about people who want to fly and have babies in space or want to fly young children before they have finished growing. The absence of gravity and the radiation environment and other things can possibly influence the growing human body.

Will it be possible for someone with a chronic disease like diabetes to visit Mars? Oh, inevitably, but not on the first expeditions. They will go as tourists, or they will go when their presence is so important that we will make the appropriate compensations for their medical condition. If there's a need for a world-famous Nobel Prize astrophysicist to go to Mars, and they happen to be a diabetic, then I am sure we will provide enough insulin and other things. Right now NASA is focusing on what it is going to take to send the first crew to Mars and the second crew and the third, not the hundredth. And we aren't in the business of colonizing Mars. We are in the business of sending people to Mars, letting them explore and come back. That will establish the basis for future colonies and future tourism and spaceflights and things like that—which are wonderful but are outside NASA's bailiwick right now.

HOW SPACE AFFECTS THE HUMAN BODY

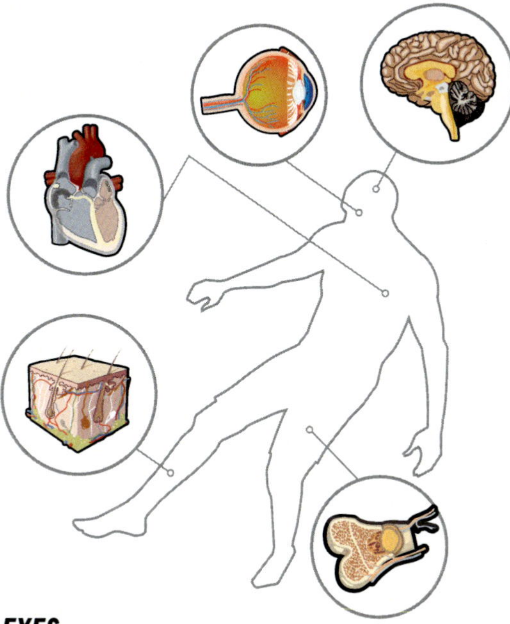

EYES
Without gravity, fluids shift upward, putting pressure on the optic nerve and eyeball. This may lead to a loss of visual acuity.

HEART
The heart muscle does not have to fight gravity to pump fluids to the upper extremities. This may cause the heart to weaken.

SKIN
Skin thins and loses elasticity. It may also become more sensitive. Wounds may take longer to heal, and infections may be more common.

BONES
Bones may lose minerals and calcium, weakening them. As calcium is flushed from the body, it may become concentrated and form kidney stones.

MUSCLES
Muscles may weaken, particularly calves, quadriceps, and neck and back muscles that support the body against the force of gravity.

EAR (BALANCE)
Weightlessness may cause motion sickness and disorientation. The body can adapt over time, but the same symptoms can return during reacclimation to gravity back on Earth.

COGNITION
Sleep may not come easily, due to tight schedules and ambient noise, leading to as little as six hours of quality sleep a night. Over long periods of time, this may cause anxiety and hinder alertness.

ROCKET BUSINESS

Entrepreneurs and their companies have taken to the skies. Can one of these commercial giants get us to Mars?

By Justin Worland

Landing a rocket on a tiny platform in the ocean is about as challenging as it sounds. The engineers and scientists at SpaceX—a commercial space-exploration company founded by billionaire entrepreneur Elon Musk—would know. They had attempted the maneuver four times, and each attempt had ended with a fiery explosion and the loss of the rocket. But in April 2016, the fifth time was the charm, as the booster stage of a Falcon 9 rocket landed squarely on a 300-foot-by-170-foot platform off the coast of Florida. SpaceX employees gathered at the company's California headquarters broke into chants of "U.S.A.! U.S.A.!" The story made international news, and office workers around the globe chatted about the topic over the watercooler for days.

The historic moment represents a key change in space exploration. Since the dawn of the Space Age, nations have competed to gain dominance over the so-called final frontier. The landing suggested that private companies could indeed deliver people and equipment to space—work long conducted by governments—in a cheaper and more efficient manner. President John F. Kennedy's famous 1961 call to put a man on the moon by the end of that decade inspired a generation of children to make "astronaut" their aspira-

tional profession. Musk's vision of colonizing Mars inspires youngsters today.

"The last few years have been transformational in a lot of ways—the emergence of a variety of different companies, capabilities and technologies is absolutely remarkable," says Eric Stallmer, president of the Commercial Spaceflight Federation. "You can say revolution or evolution, but it's one of the two."

Still, sending people and supplies to space is hard. Despite SpaceX's remarkable achievements, the company has faced setbacks. On Sept. 1, 2016, one of its Falcon 9 rockets exploded during preparations for a scheduled launch days later in an accident initially attributed by the company to an unexplained "anomaly." No one was injured, but the rocket's cargo—an Israeli Amos-6 satellite that Facebook had planned to use to bring enhanced Internet access to sub-Saharan Africa—was also destroyed. In the aftermath, many of the company's customers expressed confidence in SpaceX while not saying whether the explosion would lead them to reconsider their own launch plans. The accident is unlikely to sink the company, but it could delay Musk's ambitious plans.

Of course, NASA and the federal government are still close to the heart of space exploration. Their decades of work and research laid the foundation for private companies,

and they're still very active. But private companies—Musk's SpaceX, Jeff Bezos's Blue Origin and others—are proving essential by improving technology and building endeavors such as space tourism. "It's a natural next step," says Kathy Lueders, who heads NASA's Commercial Crew Program. "We are going to need industry smarts and investments to continue to evolve and build up the infrastructure that's needed to explore."

PRIVATE-SECTOR INVOLVEMENT IN SPACE exploration stretches back to the dawn of the space race. When Kennedy tasked NASA with that man-on-the-moon mandate, the agency turned to Boeing to build the Apollo spacecraft. The company's engineers were "the kind of people who will not permit it to fail," NASA director Robert Rowe Gilruth said at the time. Companies such as Boeing and Lockheed Martin, Northrop Grumman and their predecessors, along with hundreds of other contractors, built the satellites and shuttles that served as NASA's workhorses for the coming decades. Still, companies largely stayed away from investing their own capital in the expensive "moonshot"-type projects, instead working with federal dollars and oversight.

That changed in the 1980s with a push from President Ronald Reagan to commercialize space travel. Most important, the Commercial Space Launch Act of 1984 removed many of the regulatory barriers that discouraged private companies from going into space on their own. In the decade that followed, a flurry of companies invested in commercial space travel, but most failed to turn a profit and wound up bankrupt. (Orbital Sciences Corporation is one notable exception—it began in 1982 as a private-sector firm with no government involvement.)

In 2001, the investor and engineer Dennis Tito became the first space tourist when he paid Russian space authorities a reported $20 million for an eight-day trip to the International Space Station, thus suggesting a source of revenue unthinkable just a decade before. President George W. Bush's 2004 decision to ground the aging U.S. fleet of space shuttles less than a year after the Columbia space shuttle disaster offered the final indicator to the private sector that there was a market for commercial space products.

"NASA has been very supportive of this," says Roger D. Launius, a space historian at the National Air and Space Museum. "By guaranteeing contracts, the government has become the anchor."

Most experts will say that the most admired—and successful—private space transport company of the dozen or so in today's landscape is Musk's SpaceX. Musk, who co-founded PayPal and several other ventures, including Tesla Motors, got SpaceX under way in 2002 with eventual travels to Mars in mind. He knew, however, that the company would need to tackle less-lofty goals first. Its initial ventures focused on what is today the most profitable area of space transportation: delivering payloads to the International Space Station for the U.S. government.

"The problem with spaceflight is you literally bet the company on a single product," says Launius. "Elon is willing to spend however many hundreds of millions of his money to get SpaceX up and running."

From the beginning, SpaceX's central appeal to the federal government has been its promise to deliver reliable service at a fraction

At left, the Sept. 1, 2016, explosion of SpaceX's Falcon 9 rocket underscores the dangers of space travel; at right, one of the company's travel posters shows its promise.

SUMMIT
OLYMPUS MONS
The solar system's highest peak
MARS COLONIZATION AND TOURISM ASSOC.

BLUE ORIGIN
BLUE OR

of the price of alternatives. And that's what the company has done. SpaceX advertises the average cost of a launch of its Falcon 9 rocket at $62 million, far less than the nine figures demanded by traditional competitors such as the Boeing–Lockheed Martin collaboration United Launch Alliance. And Musk says that cost will come down further with time. Some of the cost difference can be attributed to concrete changes, such as having two rockets for a launch rather than three, as most commercial launch companies use. But the source of other cost savings is harder to discern—the company doesn't file patents or reveal design secrets. Doing things less expensively might suggest diminished reliability or safety, but SpaceX products are required to meet NASA safety standards more stringent than those of the space shuttle program before sending humans to space. SpaceX's biggest failures—like the Falcon 9 that exploded in September—did not face the same requirements manned rockets will.

In Musk's view, his competitors build products with unnecessary bells and whistles; his product gets the job done efficiently. And while old-guard government contractors tolerate costly budget overruns, Musk won't. "Is a Ferrari more reliable than a Toyota Corolla or a Honda Civic?" Musk once asked rhetorically in *Air & Space* magazine.

SpaceX has won a remarkable string of government contracts since the successful 2008 launch of Falcon 1, the first privately funded liquid-propellant rocket. The first NASA order came later that same year: 12 flights to the International Space Station, at a total cost of $1.6 billion. Since then, the company has secured contracts to launch military satellites, commercial products and, most significantly, NASA astronauts, on a mission to the International Space Station that is set to begin by the end of 2017. The total value of the contracts comes to more than $10 billion.

SpaceX competitors can be quick to point to its failures as they make the case for their own products. Bezos, the Amazon founder and CEO, seems to enjoy taunting Musk on Twitter to promote his own space company, Blue Origin. After a successful SpaceX landing in December, Bezos tweeted at Musk with apparent sarcasm: "Welcome to the club!"

Bezos has turned Blue Origin into a space powerhouse in its own right. The company's New Shepard rocket system has returned all of the rocket's constituent parts in test after test. Typically the rocket booster is dumped and lost. A reverse-propulsion system attached to the rocket returns it to the ground at a slow pace—an astounding image that makes the rocket appear as if it were taking off in reverse—and the capsule returns with parachutes slowing its landing on the ground.

But despite the accomplishments, Blue Origin has never reached orbital space, a barrier that requires more-complex systems. Unless the company can reach that level, it most likely will not win the big government contracts—though it is sufficient for the space tourism business. (Blue Origin did actually strike a deal with NASA in 2016, but that contract only provides incentives to foster innovation at the company, without a guarantee that the space agency will purchase its services.)

Unlike SpaceX, Blue Origin hopes to accomplish its goals over decades, not in a matter of just a few years. The company's slogan is *Gradatim ferociter*, Latin for "Step by step, fe-

Leading Space Companies

SPACEX

Founder: Elon Musk
Launched: 2002
Spaceships: Dragon spacecraft and Falcon 9 rocket
Pioneer: The company boasts "the first rocket completely developed in the 21st century."

BLUE ORIGIN

Founder: Jeff Bezos
Launched: 2000
Spaceship: New Shepard
Powerhouse: Financed by Bezos's wealth, Blue Origin has plenty of time to earn those big government contracts.

VIRGIN GALACTIC

Founder: Richard Branson
Launched: 2004
Spaceship: SpaceShip Two
Tourist Attraction: The Virgin company considers itself the world's first spaceline.

BOEING

Founder: William Boeing
Launched: 1916
Spaceship: CST-100 Starliner
The Original: Boeing got its start selling seaplanes to the U.S. during World War I.

rociously." And even without federal funding, Bezos has a $60 billion fortune to continue funding the company for the foreseeable future even if it does not land the lucrative government contracts to carry equipment that some of its competitors have received.

IF THE BUSINESS OF CARRYING PEOPLE and equipment to space doesn't pan out, space transport companies can turn to space tourism for cash. Blue Origin and Virgin Galactic—a venture led by British billionaire Richard Branson—have lined up hundreds of customers willing to pay a few hundred thousand dollars each for a quick ride in space. Top selling points include a stunning view of Earth at more than 300,000 feet away from the planet—an impressive height but still at the suborbital level—and a few moments of gravity-free existence.

Despite enthusiasm from celebrities and multimillionaires, space tourism has hit a few snags over the past few years. Branson suggested that the company might make its first space trip by 2009 when he unveiled the system in 2008, but a number of technical difficulties have delayed the launch. And those troubles added to a series of serious complications, including the 2014 crash of a Virgin Galactic craft that killed the pilot during a test run. The combination of delays and disturbing safety-related questions following the crash led many would-be space tourists to withdraw.

But for committed billionaires like Branson, crashes act only as minor setbacks. Branson flew to the site of the crash and promised that safety would continue to be a top priority. But he also responded defiantly and promised to keep building.

The true achievement for space tourism may be launching tourists beyond the suborbital level. Unsurprisingly, SpaceX has suggested that it may eventually use its equipment for that purpose. But for now, that must remain the domain of the super-rich willing to pay many millions to the Russian space program—though they need to wait until 2018, when the program restarts after a hiatus. A total of eight space tourists were launched between 2001 and 2009.

Delivering passengers and cargo to space

Above: The New Shepard booster launches—and also lands—vertically. Right: SpaceX tested Dragon's abort system in 2015. Below: Richard Branson with Virgin Galactic's SpaceShipTwo.

is enough to excite millions of space observers, but today's entrepreneurs see it as just the beginning of galactic possibilities. SpaceX's Red Dragon program, scheduled to launch in 2018, would be a key development in Musk's plan to take humans to Mars. The Red Dragon would greatly simplify how humans approach Mars—and reduce the cost. Musk's plan is to lower the landing craft to Mars with the same technology, called propulsive landing, that lowers it to Earth in the much-famed Atlantic landings. The Dragon capsule would help test Mars landings for future missions and allow the company to explore the planet's surface. The launch will cost $90 million, Musk says.

All of this Mars exploration is meant primarily to prepare for future human exploration of the planet—or at least that's what Musk says. He has even discussed building a million-person colony where people settle for good. Musk's insistence on getting started in 2018 is in part his exuberance for achieving the seemingly impossible, and it also necessarily coincides with a time when Mars will be closest to Earth—a period that occurs approximately every two years. Musk's further timeline, of seeing actual human exploration of Mars within a decade, causes skepticism among many space experts. "Now is the first time in the history of Earth that the window is open, where it's possible for us to extend life to another planet," Musk said in late 2015. "I think the wise move is to make life multiplanetary while we can."

Private enterprise would most likely have a role to play in any future colonization or even human exploration of Mars. Indeed, NASA has entered into a "no exchange of funds" agreement for Mars exploration, signaling that the agency wants to help private companies. The agreement means that companies and NASA will share resources—mostly research and know-how—without demanding anything in return except the hope that the collaboration will eventually lead us to Mars.

The growth of revolutionary companies— Apple, Netflix, Southwest Airlines and many others—typically comes at the detriment of more-established industry players. That's also the case in the world of space exploration. Companies such as Boeing, Northrop Grumman and Lockheed Martin have long enjoyed the advantage of being the only companies able to construct spacecraft that fit the stringent requirements of the federal government. And with that semi-monopoly, the companies have been able to charge prices that provide high profit margins and feed that income back to their shareholders. NASA and the federal government raised little complaint about that system for decades. Indeed, the savings of a few million—or even $100 million—made scant difference to an agency that was once awash in federal funds. Part of that inclination to favor reliability rather than taking risks explains why NASA initially approached some private space ventures with skepticism.

But today NASA officials have come around to saying the new system is better than before, fostering innovation that will drive the next century of space exploration. That's part of what Kathy Lueders says she brings to her commercial crew leadership at NASA: she has to consider not only whether a proposal will serve her agency's specific goal but also whether funding it will pay dividends in other areas. NASA's COTS program, short for Commercial Orbital Transportation Services, aims to do just that. The federal government provides funding to promising companies when they meet certain advances to help facilitate their growth. SpaceX and Blue Origin were both recipients of that funding. All these changes drive a transition of space into a commercial area that resembles the transition of aviation and telecommunications, Lueders says. And with the minds of some of the world's best innovators working on it, the possibilities are endless.

"I don't know where we're going yet," says Lueders, "but I know that when we look back at this time 50 years from now, we'll recognize this as a renaissance of growth of knowledge and capability."

> ## "WHEN WE LOOK BACK, WE WILL RECOGNIZE THIS AS A RENAISSANCE OF GROWTH."

LOCAL COLOR
The pits of impact craters
such as this one in the ancient
Martian highlands often expose
deep bedrock of many colors. A
2012 study mapped more than
600,000 such craters on Mars,
caused by meteorites.

CHAPTER THREE
THE ALLURE
MARS DRAWS
US IN WITH ITS
POTENTIAL AND
ITS MYSTIQUE

MARS ART

Ever since the first close-up image of Mars in 1965, NASA scientists have used pictures of the planet to gain insight into its geography. These landscape images are imbued with false colors to show a range of variations that would not be noticeable in a single color. The result: Martian landscapes that look like they've sprung from a painter's palette

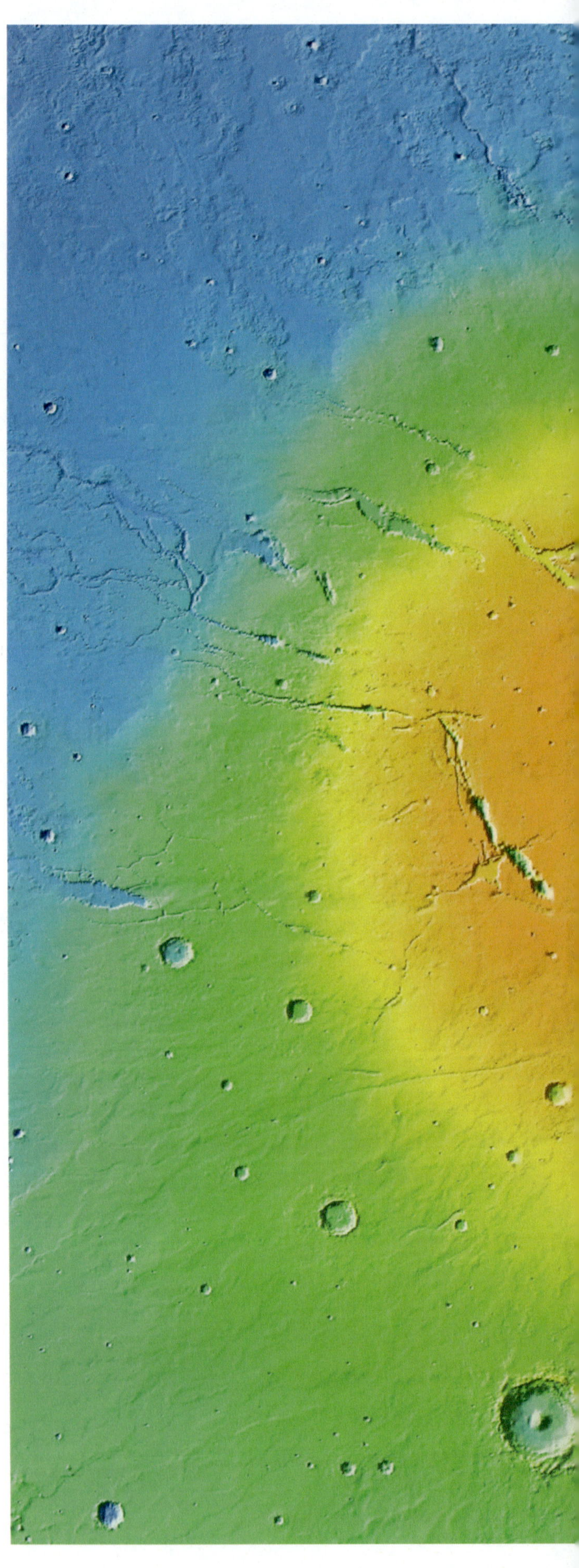

VIBRANT ERUPTIONS

These crimson mounds are three of the volcanoes that make up Elysium Planitia, the second-largest volcanic region on the planet. The largest volcano, Elysium Mons, is at the center and rises eight miles above the surface.

A NEW FORMATION

The impact crater above stretches 100 feet. NASA's Mars Reconnaissance Orbiter observed the first appearance of this crater in May 2012. It was not captured during the July 2010 observation, meaning that sometime between those two dates, this crater was formed.

BUILT BY THE BREEZE

Wind picks up debris and sculpts ridges like those at left. These erosions are called yardings, and because of Mars's thin atmosphere, the shadows cast are darker than those on Earth.

GLOWING GULLIES

The five-mile-wide crater at right contains several gullies. NASA ran tests to determine if water could have caused the slopes. While water's presence here was not ruled out, the results confirmed that dry materials, like sand, created these gullies.

A CRATER NAMED GALE

About 3.5 billion years ago, a meteor crashed into Mars and created the 96-mile cavity below. The Gale Crater houses Mount Sharp, which stretches three miles up from the crater's floor.

SLIDING INTO "HOME PLATE"

Named for its similarity to bases on a ball field, the rock outcrop above is housed in Gusev Crater. Home Plate's debris was caused by a hydrovolcanic explosion, in which molten basalt from an erupting volcano comes into contact with liquid below the surface.

ERUPTIONS

Lava from the vast volcanic region of Tharsis found its way to the cratered landscape of Terra Sirenum, where scientists say the relationship between its flows and craters tells a complex tale.

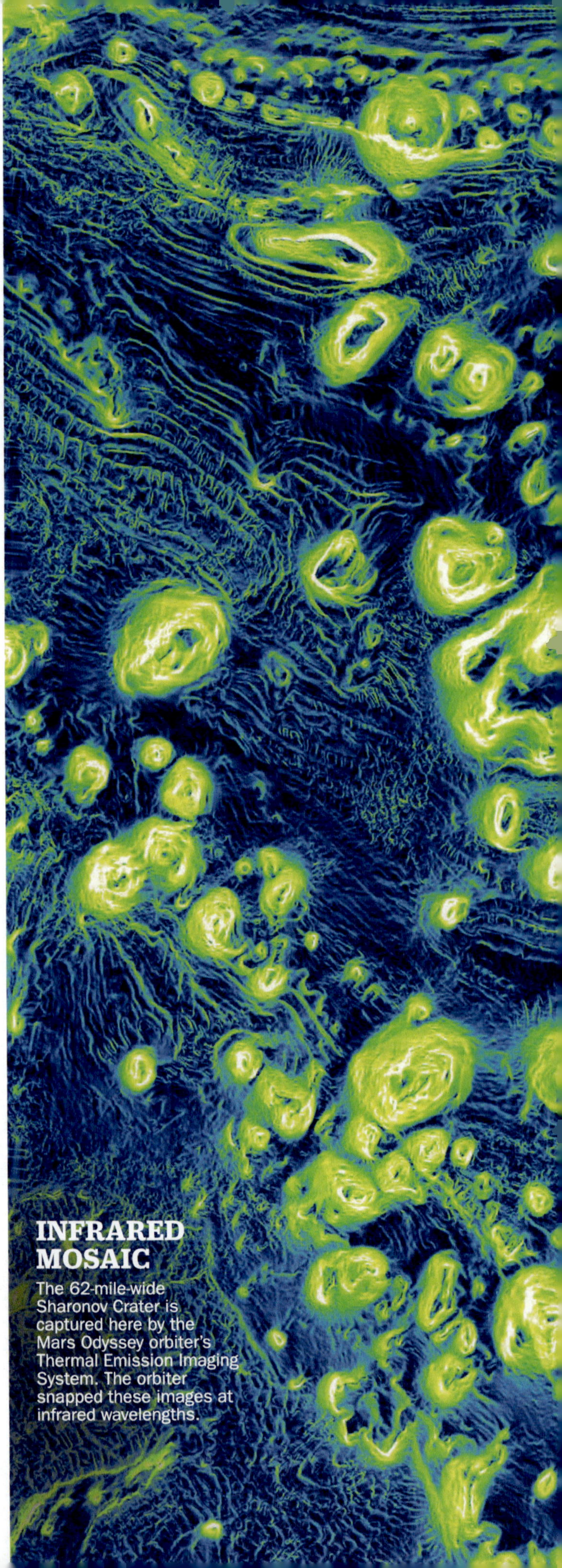

INFRARED MOSAIC

The 62-mile-wide Sharonov Crater is captured here by the Mars Odyssey orbiter's Thermal Emission Imaging System. The orbiter snapped these images at infrared wavelengths.

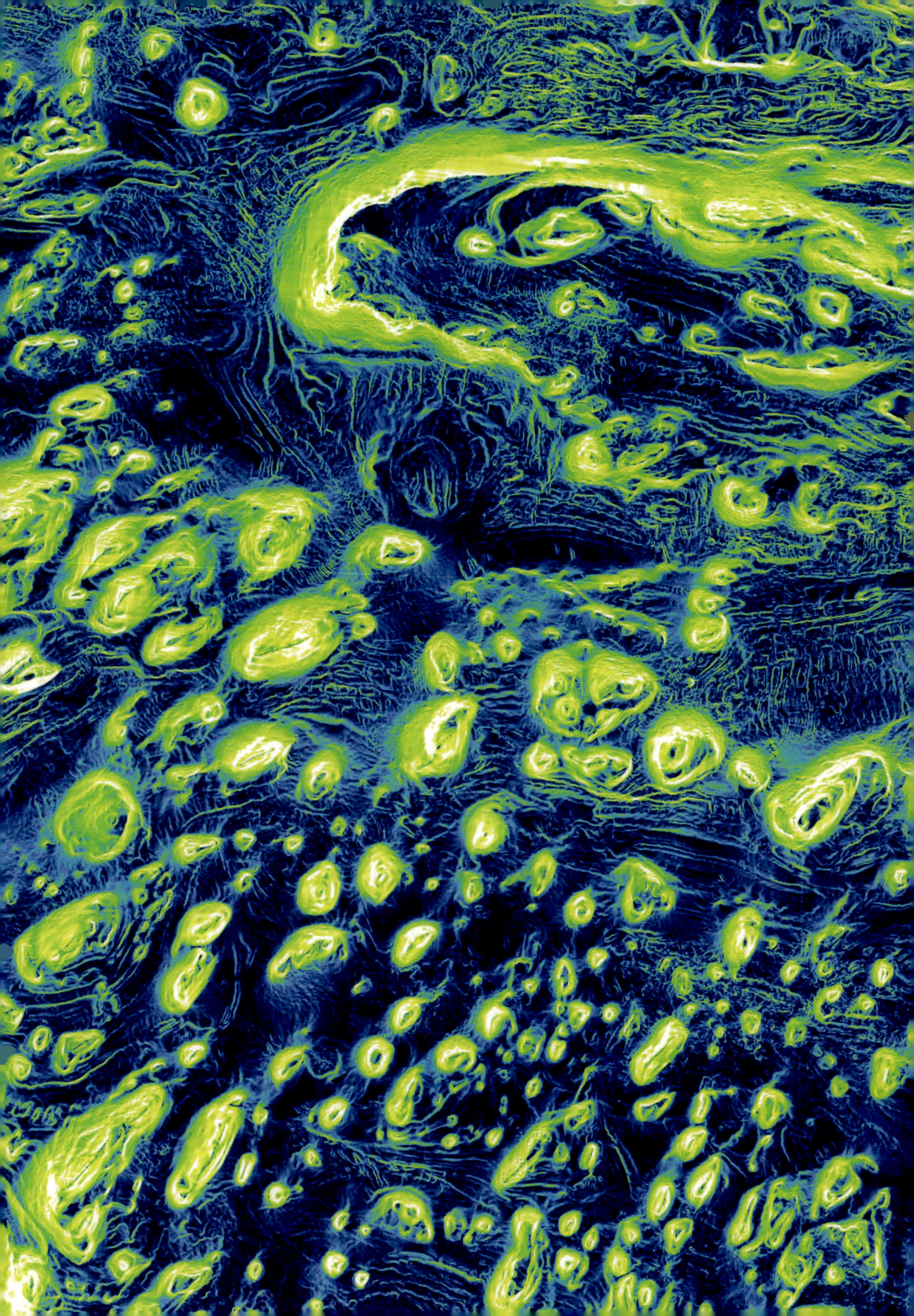

THE RED MUSE

The planet inspires the imaginations of writers, musicians, animators and, of course, filmmakers

By Daniel D'Addario

It's a strange feeling," Mark Watney says at one point during his lengthy stay on Mars. "Everywhere I go, I'm the first. Step outside the rover? First guy ever to be there! Climb a hill? First guy to climb that hill! Kick a rock? That rock hadn't moved in a million years!" In the 2015 blockbuster film *The Martian*, Watney (played by Matt Damon) is half exhilarated and half distressed by his unique situation. After a NASA expedition goes awry, he's marooned on a planet that's uncannily like Earth in a few ways—but missing key components. If you're a brilliant botanist, as Watney is, you can grow potatoes on the planet. But survival isn't just lonely: it's grueling, thanks to Mars's rocky landscape and unfamiliar, inhospitable atmosphere and soil. The movie's conclusion is triumphant because Mars does not want Watney to survive.

The soil of Mars may have made rudimentary potato farming a challenge, but it's got the stuff to make stories blossom. This is a planet

Matt Damon stars as the only man on Mars in the 2015 hit movie *The Martian*.

Edgar Rice Burroughs, above, published *A Princess of Mars*, the first novel in his Barsoom series, in 1917. A movie adaptation, *John Carter*, starring Taylor Kitsch and Lynn Collins, right, was released in 2012.

with days that last about the same length as Earth's and which is made up of similar rocky stuff, but it's one that can't keep us alive. Its distinctive, eerie red soil couldn't have been made up by the most inventive Gothic novelist; the universe provided almost too obvious a symbol of hostility.

While *The Martian*, based on a novel by computer programmer Andy Weir, accurately represents the threats Mars posed, earlier renderings of the planet tended to favor a more impressionistic tone. Consider the work of Edgar Rice Burroughs. The Tarzan creator was beloved for his romantic depiction of Mars in the Barsoom series, first published in serial form in 1912. Burroughs's Mars, visited by Virginian soldier John Carter, is populated by alluring and advanced species, like that of Dejah Thoris, the red-skinned woman who gives Burroughs's novel *A Princess of Mars* its title.

Burroughs's imagining of Mars had little to do with the planet's specificities, so much so that he gives it a new name. "Barsoom" is what native Martians call Mars—the first sign that Carter is a true outsider. Carter had been transported to Mars by a mysterious portal after fighting in the Civil War. The low gravity

and thin atmosphere, in this imaginative telling, aren't an impediment to life but the very things that allow a man accustomed to Earth's physics to have the agility and strength of a superhero. In Burroughs's telling, Mars is the next place to conquer.

Burroughs's series, which today comprises 11 books published from the 1910s to after the author's death in 1950, allowed John Carter to prove his inherent greatness simply by taking a step. By midcentury, though, the breathless tone of mainstream science fiction and fantasy gave way to something a bit more complicated. Ray Bradbury's 1950 *The Martian Chronicles*, for instance, is among the best-remembered works by the sci-fi master. The short-story collection depicts humanity attempting to colonize the Red Planet while bringing with it all of its faults.

As in Burroughs's work, Bradbury's Mars seems in many ways a refuge from earthly danger. John Carter finds himself on the Red Planet as he's evading capture by Apaches in the American Southwest. Bradbury's humans end up on Mars for reasons reflective of the author's own time. During the Cold War, with all its tensions over the potential use of nuclear

Ray Bradbury's classic short-story collection *The Martian Chronicles* was published in 1950. Roddy McDowall, right, stars in the *Twilight Zone* episode "People Are Alike All Over" in 1960.

weapons, Bradbury's depiction of an Earth so threatened by the possibility of war that space represents a great escape must have seemed particularly relatable. The American "Red Scare," which targeted perceived Communist agents among the intellectual class, made its way into Bradbury's work, too: one astronaut is a man of letters seeking safe haven after the government destroyed his book collection. (Bradbury went on to probe similar concerns in 1953's *Fahrenheit 451*.)

Bradbury's Mars is far from an easy place. Its natives are understandably unwilling to cede their planet to humans, whose cruelty Bradbury is unafraid to depict. (In one memorable instance, young human boys play the bones of dead Martians like "white xylophones," desecrating the remains of the planet's former inhabitants as their parents construct a replica of American life.) Here, Mars, with its close-enough-to-Earth aspect and its attractive proximity to our planet, becomes an opportunity for an imaginative writer to assert that sometimes we are the aliens.

That outlook was shared by *The Twilight Zone*, the pioneering science-fiction TV series that pursued philosophical questions with rel-ish. The first-season episode "People Are Alike All Over," which aired in 1960, places an optimistic astronaut's belief that Mars's inhabitants will be friendly and benevolent—just like humans!—in contrast with his spacecraft's crash landing. The Martians are indeed just like humans: they put the earthling in captivity and treat him like a zoo animal. Other *Twilight Zone* episodes treat the planet as a threat to our way of life, with Martians conducting degrading experiments on people or laying the groundwork to construct colonies. The planet is shorthand for a threat of invasion just close enough to disturb the viewer's sleep. But what "People Are Alike All Over" suggests is that, if real, Martians might simply be as callous in expressing interest in our way of life as we would be, were the situation reversed.

WHILE ADULTS WERE READING *THE MARtian Chronicles* and watching *The Twilight Zone*, their children were getting a vision of Martian life that tempered its threat of violence with humor. Looney Tunes' Marvin the Martian, with a centurion's helmet and an inscrutable, big-eyed face, has since 1948 sought to destroy Earth. Our planet, he perpetually frets, obstructs his view of Venus. In his first appearance, Marvin is on the moon, attempting to blow up Earth. Bugs Bunny, on his own space mission, outwits Marvin, and not for the last time. One of Marvin's catchphrases is "Where's the kaboom?," uttered whenever his latest attempt to blow up our planet falls short.

Marvin is as reliable a foil as Elmer Fudd or Yosemite Sam. Every time he comes back, it is with new optimism that he'll defeat the ever-resourceful Bugs. And each time, Bugs wins out, though sometimes with collateral damage. The 1958 short "Hare-Way to the Stars" ends with Bugs inadvertently bringing a horde of dehydrated Martians to Earth and soaking them in water. The conquering horde of aliens is forgotten by the time the next Bugs Bunny adventure starts.

That possibility—of inflicting damage beyond the welfare of one Brooklyn-accented rabbit—makes Marvin a bit more exciting than other Looney Tunes villains. Though his easily roused frustration, so unlike Yosemite Sam's brashness, makes him sweetly relatable, Marvin winning out in even one episode would mean the end of the world. He is, like *The Twilight Zone*'s Martians, recognizably human but as anti-human as can be.

At the same time that humanity was gearing up to, for the first time, explore space, different entertainments pitched at adults and at kids were warning of malevolence originating from Mars. There are far fewer depictions of, say, Venusians in culture (though they were occasionally used on *The Twilight Zone* as rivals to Martians); it's hard to call to mind any inhabitants of Mercury or Neptune. Depicting life on Mars—even in as safe a way as through a perpetually thwarted character in a children's cartoon—harnesses the extremity of space's possibility in a way that's literally close to home. The Americans of *The Martian Chronicles* can crate all their belongings and customs to Mars without having to go into hypersleep; Marvin is both completely freakish and acquainted enough with Earth that he and Bugs can speak a common language.

That kind of Martian paradox was shared by David Bowie, who could both create viscerally danceable music and appear to have transcended ordinary humanness. He sought

Marvin the Martian and Bugs Bunny, above, have dueled for decades in Looney Tunes cartoons. David Bowie, right, as his alter ego Ziggy Stardust, in 1972. Below, Arnold Schwarzenegger stars in *Total Recall*, 1990.

to appear at once like a man and an extraterrestrial, and he chose Mars as a sort of totem with his 1971 single "Life on Mars." The song, one of Bowie's signature hits, uses bizarre imagery—of a caveman, of police violence, of Mickey Mouse having "grown up a cow"—to show just how grim and dull life on Earth has become. As Bowie sings, it's "a saddening bore." The repeated question "Is there life on Mars?" feels alternately plaintive and hopeful. The answer would seem to be no, but Bowie's extraplanetary inventiveness made it seem, at least for the length of a song, possible.

And that was before his 1972 album *The Rise and Fall of Ziggy Stardust and the Spiders from Mars*, on which Bowie represented himself as Ziggy, the messenger of "Starman," a creature from out there in the galaxy. "He'd like to come and meet us," we're told, "but he thinks he'd blow our minds." It fell to Bowie to interpret his messages for us. Mars stood in for every part of the universe that felt undiscovered, pure and filled with potential. Stuck on Earth, with all the dullness that implies, Ziggy sustains himself through the adoration of his fans and through prodigious excess. Space didn't have to feel like a last resort before humanity destroyed itself or like an existential threat. At least not for Bowie. In an era of renewed optimism about what might lie elsewhere in our solar system, Bowie suggested that Mars represented possibilities for something other than conquest or being conquered.

BOWIE'S TONE IS REFRESHINGLY ORIGINAL among the many popular associations with space as the site for horror and action. The 1987 graphic-novel classic *Watchmen* movingly depicts Mars as the place of exile for a superhero unwilling to subject Earth to his godlike powers. Its desert landscape suits his mood of desolation. But more emblematic of our attitude toward our neighbor is 1990's *Total Recall*, depicting a future in which Mars is the domain of rapacious, evil humans. Just as in *The Martian Chronicles*, humanity doesn't change its nature just because it changed its location. Here, Arnold Schwarzenegger fights off conspirators

who seek to keep a monopoly on Mars's oxygen. The 1993 video game *Doom*, which has generated sequels and a feature film, depicts a Marine stationed on Mars to guard a mega-corporation's scientific experiments and face down the planet's monsters. Here, Mars is both humanity's future and the site of its destruction.

That double understanding of Mars has continued. In 2000, two films about humanity's pursuit of a new home on Mars came out. *Red Planet* depicts Mars as nearly impossible to survive on but filled with potential. Brian De Palma's *Mission to Mars* is, fitting its director's background in psychological horror, odder yet, with a Mars that seems to be actively fighting back against its potential colonists. We can't stop going there on film, even as those films acknowledge that going to Mars is a very bad idea.

The public seemed, for a while, to have turned on Mars as a subject of inquiry. *Mars Needs Moms*, a 2011 animated film depicting a child's attempt to rescue his abducted mother, was a very expensive box-office bomb. The same was true of *John Carter*, a 2012 film adapted from Burroughs's stories of Barsoom. Those tales' sense of giddy adventurism, perhaps, had come to feel a bit out of vogue after decades' worth of tales that treated space as dangerous or at least treated it seriously.

Fantasies of Mars, from *The Twilight Zone*'s imagining of a race as small-minded as our own to Bowie's metaphorical invocation of its extraordinary possibilities, have provided years of fear and fun. *The Martian* is the perfect movie for an era in which we know more about Mars than ever before. Colored by scientific knowledge, the movie shows Mars as both eerily beautiful and as itself: a hunk of rock. It is an unspoiled blank slate, a desert that has no character save for what the Martian brings to it. He's the first man to have kicked its rocks; his story is the first to imagine the planet in such a nourishingly scientific way, bolstered by facts to which Bradbury and Burroughs had no access. Long may Mars's emptiness, and the way it hangs just close enough in the sky, inspire writers and filmmakers to bring to it their own firsts.

> # MARS STOOD IN FOR EVERY PART OF THE UNIVERSE THAT FELT UNDISCOVERED.

WILL MAN OUTGROW THE EARTH?

In a 1952 cover story, TIME predicted how we might get to Mars

The youngsters have already zoomed confidently off into the vast ocean of space; they can buy space suits, space guns and rockets in almost any toyshop. In 50-odd science fiction magazines, space travel is a favorite theme. Eight comic strips and at least two TV programs are flying through space. "Scientific" space books are brisk sellers. But not all members of the space cult are storytellers, crackpots or kids. Some serious scientists believe that space flight will surely come, and perhaps soon, but they know that separating facts and fancy about space travel is almost as difficult as a trip to the moon.

The basic scientific principles were worked out long before World War I, but the popular vogue for space travel probably grew out of two great technical achievements of World War II. Nuclear fission convinced the public that "science can do anything." The German V-2 rocket, the world's first guided long-range missile, proved that a man-made vehicle can climb into space, however briefly.

Like Columbus?

Space enthusiasts like to compare the present with the time just before Columbus, when Europeans were about ready to launch out over the Atlantic. The analogy is poor. Columbus did not know what he would find on the other side of the ocean, but he had ships that would take him across. The space men can see across their "ocean," but they have no ships. Today, the Cold War has thrown a blackout over all rocket research, such that rocket experts, working on guided missiles, are sworn to secrecy. Not one man on earth who knows the latest developments can talk freely about them. Men who do not know can let their fancies run wild, for they have no fear of expert contradiction.

Up from Gravity

The best way to visualize space in terms of astronavigation is to think of it as a placid lake with a few widely separated whirlpools in its mirror surface. These sucking danger spots are the gravitational fields around the sun and its satellites. The cardinal principle of astronavigation is to keep far away from gravitational maelstroms. Unfortunately for the space men, ships will have to set sail from the middle of one: the strong gravitational field that surrounds the earth.

The energy needed to escape from the earth's suction is simple for astronauts to figure. Expressed as speed, it is 25,000 m.p.h. A spaceship with this "escape velocity" would be an independent part of the solar system and could cruise, with a little more energy, all over the place.

Twenty-five thousand miles an hour is faster than a single rocket can travel, but long before World War II, the space men thought of a trick: the multi-stage rocket. This is a "beast" that shoots upward with a smaller beast attached to its nose, and as it travels it sheds its parts in what are called "stages." The trouble is, each stage must be enormously larger than the next stage. Rocket men argue endlessly about the details, but the more sensible ones believe that it would take a multistage rocket as big as an ocean liner to spit even a jeep-sized spaceship free of the earth.

The space men believe that the leap from the earth must be made in two jumps, with a resting and refueling spot partway up the slope of the earth's gravitational whirlpool. If a rocket is shot straight up at less than escape velocity, the rate and speed needed to escape the earth's gravitational pull, and makes the proper turn as it clears the atmosphere, it will be sidetracked to an orbit above the earth and will circle around endlessly. The centrifugal force of the rocket's motion around the earth will exactly balance the pull of the earth's gravitation. This same balance of forces keeps the moon on its rails.

Since it takes less energy to reach an orbit than to escape from the earth, astronauts believe that a moderate-sized three-stage rocket, or even a two-stage one, could make the trip with a good payload. The rocket would park its load (e.g., fuel) in the orbit, where it would circle as safely as if it were back at the filling station.

After enough fuel had been delivered to the little, man-made satellite, a rocket could fill its tanks and blast itself off into space. Since it would already be moving in its earth-circling orbit at a good clip (16,000 m.p.h.), the rocket would need only a moderate additional push to give it escape velocity. Then it could cruise freely in space, like a ship that has risen out of a whirlpool and reached the smooth surface of a lake.

Approach with Caution

Plans for space travel may sound dull, but astronauts already are planning more exciting jaunts. One plan for a trip to Venus, for instance, uses space ships from an orbit around the earth to establish a base on the moon. A special ship then takes off from the moon at a moment when Venus is considerably behind both earth and moon on its shorter and faster orbit around the sun.

Larger bodies, such as Mars and Venus, both powerful gravitational whirlpools, should be approached with caution. But the planets both have atmospheres, which the space men plan to use as frictional buffers. Their ships would circle in the atmospheric fringes until they were moving slowly enough to land. An alternate plan: cruise warily around the planet and send small spacedinghies down to explore its surface.

It's important to note that the first step toward space travel should be to set up a special commission to study the entire matter. Its members should be scientists, engineers and economists of the highest type. They should be "cleared" to receive all the latest news of guided missile progress, and they should be above interservice rivalries and the self-seeking pressure of missile manufacturers. After making their decision on the feasibility and value of all types of space vehicles, they should lay down a practical program for the U.S. to follow.

The public will not be told the decision of such a commission. To confirm its formation or to deny it would reveal the summation of many military secrets. If the commission is a go, the first news for the public may be an "American star" [meaning a spacecraft], rising in the west and sweeping swiftly across the sky.

FACT OR FICTION: HOW WELL DO YOU KNOW MARS?

Truths and falsehoods about the Red Planet

By Nicole Fisher

1. Just as in *The Martian*, a violent dust storm could leave an astronaut stranded on the planet.
FALSE With less atmospheric pressure than Earth, Mars has winds that top out at 60 miles per hour. That is less than half the speed of the worst hurricane winds we experience on Earth.

2. Mars could develop rings like Saturn.
TRUE Phobos, the smaller Martian moon, is gradually spiraling inward. It will eventually be so close that the planet's gravitational pull could tear the moon apart, creating a ring. But don't wait up: that could take 20 million to 40 million years.

3. On August 27 of each year, Mars appears, to the naked eye, to be the same size as the moon.
FALSE The Mars Hoax started as an email chain in 2002 and has resurfaced every year since. While Mars's orbit was only 35 million miles away in 2003, the closest in recorded history, observers would have still needed a telescope to make it appear as large as the moon.

Saturday Night Live's Coneheads were out of this world.

4. *The Twilight Zone* describes Mars as being 11 million miles away. Is this right?
FALSE Depending on the two planets' positions in their orbits, Mars is an average of 140 million miles from Earth.

5. Astronauts could one day visit Venus.
TRUE Kind of. Engineers and scientists have been working on HAVOC, or High Altitude Venus Operational Concept. This preliminary study involves a theoretical city floating in Venus's atmosphere, where conditions are more suitable for life than on the planet's scorching surface. A trip to the atmosphere around Venus, our closest planet on average, could be a test run for a Mars mission.

6. There's a volcano on Mars that dwarfs Mount Everest.
TRUE Olympus Mons on Mars is the tallest known volcano in the solar system, at 13.6 miles high. Everest is a mere 5.5 miles.

7. The Coneheads are from Mars.
FALSE The iconic *Saturday Night Live* characters descended here from the planet Remulak.

8. Mars is private property.
FALSE When the Sojourner rover touched down on Mars in 1997, three (perhaps a bit spacey) Yemenite men sued NASA for trespassing, asserting that they inherited the planet from their ancestors 3,000 years ago. Luckily for NASA, 1967's multinational Outer Space Treaty had declared Mars the property of all humanity.

9. Sunsets on Mars appear blue.
TRUE The fine dust in Mars's atmosphere allows light on the blue end of the spectrum to penetrate more efficiently than longer-wavelength colors. When the air is thickest, near sunrise and sunset, the blue light is more pronounced. You can see the blue glow above the sunset on the back cover image.

TIME

Editor Nancy Gibbs
Creative Director D.W. Pine
Director of Photography Kira Pollack

MISSION TO MARS: OUR JOURNEY CONTINUES

Editors Kostya Kennedy, Courtney Mifsud
Designer Skye Gurney
Photo Editor Crary Pullen
Writers Jeffrey Kluger, Buzz Aldrin, Daniel D'Addario, Lisa Eadicicco, Nicole Fisher, Lily Rothman, Alexandra Sifferlin, Justin Worland
Reporters Elizabeth L. Bland, Emily Barone
Editorial Production David Sloan

TIME INC. BOOKS
Publisher Margot Schupf
Associate Publisher Allison Devlin
Vice President, Finance Terri Lombardi
Vice President, Marketing Jeremy Biloon
Executive Director, Marketing Services Carol Pittard
Director, Brand Marketing Jean Kennedy
Finance Director Kevin Harrington
Assistant General Counsel Andrew Goldberg
Assistant Director, Production Susan Chodakiewicz
Senior Manager, Category Marketing Bryan Christian
Brand Manager Katherine Barnet
Associate Prepress Manager Alex Voznesenskiy
Project Manager Hillary Leary

Editorial Director Kostya Kennedy
Creative Director Gary Stewart
Director of Photography Christina Lieberman
Editorial Operations Director Jamie Roth Major
Senior Editor Alyssa Smith
Assistant Art Director Anne-Michelle Gallero
Copy Chief Rina Bander
Assistant Managing Editor Gina Scauzillo
Assistant Editor Courtney Mifsud
Special thanks: Don Armstrong, Nicole Fisher, Kristina Jutzi, Seniqua Koger, Kate Roncinske

This close-up of sand dunes on Mars's northern polar cap shows patches of ice during the Martian summer.

CREDITS

FRONT COVER
Photo illustration for TIME: (Mars) NASA/JPL-Caltech/USGS; (background) Detlev van Ravenswaay/Getty Images

BACK COVER
NASA/JPL/Texas A&M/Cornell

TITLE PAGE ESA/ATG medialab

CONTENTS
2–3 NASA/Marshall Space Flight Center

INTRODUCTION
4–5 NASA/JPL-Caltech/MSSS **6–7** (top) From The New York Times (December 9, 1906) © 1906 The New York Times; (bottom) NASA/JPL-Caltech/MSSS

FOREWORD
11 NASA **12** (from left) NASA; Katherine Frey/The Washington Post/Getty Images

THE JOURNEY
14–15 NASA/JPL-Caltech/Arizona State University **16–17** Bill Ingalls/NASA **18–19** NASA/JPL-Caltech/MSSS **20** NASA/JPL-Caltech/Malin Space Science Systems **23** Abhishek N. Chinnappa/Reuters **24–25** Lon Tweeten for TIME **27** NASA **29** (clockwise from top right) public domain/Wikimedia Commons; Bettmann Archive/Getty Images; Paul Slade/Paris Match/Getty Images; Stephen Jaffee/AFP/Getty Images; NASA/National Archives; NASA; Cecil Stoughton/John F. Kennedy Library and Museum, Boston **31** NASA/JPL-Caltech/MSSS **32–33** (clockwise from top left) NASA; NASA/JPL-Caltech/Space Science Institute; NASA; NASA/JPL (2); NASA (2) **34–35** (clockwise from top left) NASA; NASA/JPL (2); Nesnad/Wikipedia; NASA/JPL (3) **36–37** Lon Tweeten for TIME **38–39** NASA/Johns Hopkins University Applied Physics Laboratory/

Carnegie Institution of Washington **40** NASA/JPL **41** Damian Peach/NASA **42** NASA/JPL **43** NASA/JPL/Caltech/Space Science Institute **44** NASA/JPL **45** NASA/Johns Hopkins University Applied Physics Laboratory/Southwest Research Institute **46** NASA/JPL-Caltech/SETI Institute **47** NASA/JPL/Space Science Institute

THE PLAN
48–49 NASA/JPL-Caltech/University of Arizona **50–51** NASA/Kennedy **52** Bill Ingalls/NASA **53** NASA/MSFC **55** (from top) David C. Bowman/NASA, Langley; NASA, Langley; David C. Bowman/NASA, Langley **56** Win McNamee/Getty Images **58–59** (clockwise from top left) Ross Lockwood; Oleg Abramov; Sian Proctor; Neil Scheibelhut/HI-SEAS, University of Hawaii (4) **60–63** Bill Ingalls/NASA (2) **65** Scott Kelly/NASA **67** James Blair/NASA/JSC **69** Heather Jones for TIME **71** Patrick Fallon **72** USLaunchReport **73** Courtesy SpaceX **74** Courtesy Blue Origin **76** (from top) courtesy Blue Origin; courtesy Ben Cooper/SpaceX; Mark J. Terrill/AP

THE ALLURE
78–79 NASA/JPL-Caltech/University of Arizona/HI-RISE **80–82** NASA/JPL-Caltech/Arizona State University (3) **83** NASA/JPL-Caltech/University of Arizona/HI-RISE (2) **84** (from top) NASA/JPL-Caltech/USGS/Cornell University; NASA/JPL-Caltech/University of Arizona **85** NASA/JPL-Caltech/University of Arizona **86–87** 20th Century Fox/Everett Collection **88** (from left) Chicago History Museum/Getty Images; Walt Disney Company/Everett Collection **89** (from left) courtesy Doubleday; CBS via Getty Images **90** (from top) Everett Collection; Michael Ochs/Getty Images; The Kobal Collection **92** Courtesy of Time Inc. **94** NBC/NBCU Photo Bank/Getty Images **95** NASA/JPL-Caltech/University of Arizona **96** Bettmann Archive/Getty Images

MARS
REVOLUTIONARIES
It was through this
behemoth of a
telescope that a Naval
Observatory professor,
Asaph Hall, discovered
the two moons of Mars,
Phobos and Deimos, in
1877. Here, his son,
also named Asaph, has
a look-see in 1924.